MUSHROOM CULTURE

Its Extension and Improvement

MJP
PUBLISHERS

MUSHROOM CULTURE

Its Extension and Improvement

W. ROBINSON

MJP PUBLISHERS

Chennai New Delhi Tirunelveli

ISBN 978-81-8094-343-0

MJP Publishers
No. 44, Nallathambi Street,
Triplicane, Chennai- 600 005

Printed and bound in India

MJP 395

Publisher : C. Janarthanan

PUBLISHER'S NOTE

The legacy of a country is in its varied cultural heritage, historical literature, developments in the field of economy and science. The top nations in the world are competing in the field of science, economy and literature. This vast legacy has to be conserved and documented so that it can be bestowed to the future generation. The knowledge of this legacy is slowly getting perished in the present generation due to lack of documentation.

Keeping this in mind, the concern with retrospective acquiring of rare books has been accented recently by the burgeoning reprint industry. MJP Publishers is gratified to retrieve the rare collections with a view to bring back those books that were landmarks in their time.

In this effort, a series of rare books would be republished under the banner, "MJP Publishers". The books in the reprint series have been carefully selected for their contemporary usefulness as well as their historical importance within the intellectual. We reconstruct the book with slight enhancements made for better presentation, without affecting the contents of the original edition.

Most of the works selected for republishing covers a huge range of subjects, from history to anthropology. We believe this reprint edition will be a service to the numerous researchers and practitioners active in this fascinating field. We allow readers to experience the wonder of peering into a scholarly work of the highest order and seminal significance.

MJP Publishers

PREFACE

MY reasons for writing this book are: First, that Mushroom Culture is but little practised in this country compared to the extent to which it ought to be, considering the abundance of the necessary materials in all parts of these islands, both in town and country, and the high estimation in which the Mushroom is held. I now refer to ordinary Mushroom Culture as practised in our best private gardens. I believe it possible and desirable to extend this, the only phase of the Culture that can be called popular, in a tenfold degree, and that every place in which a gardener and horses are kept should be abundantly supplied with Mushrooms throughout the greater part of the year. Secondly, that although Mushroom Culture as usually practised is perfectly well known to good cultivators, a simpler and fuller account of it than has yet appeared in any English book on the subject is desirable for the unpractised amateur and cultivator. Thirdly, that Mushroom Culture is at present confined to a too narrow groove; and a belief that the general gardening public should have a broad and clear idea of the several ways in which they may procure abundance of excellent Mushrooms with very trifling expense. Even many of the best private growers never think of it except as illustrated on their comparatively small beds in small houses. I believe that if the knowledge of how easily and in how many ways they may be grown, apart from the usual mode, were sufficiently spread, it would lead to the production of many times our present supply. Fourthly, a desire to introduce to this and other countries the system of Mushroom Culture on a very large scale carried on in caverns beneath the environs of Paris, which caverns I visited in 1868.

To these reasons I might add a wish to call attention to the waste of money for Mushroom-spawn that now occurs in nearly every garden. There is not the slightest necessity for this. In every garden where Mushrooms are grown abundance of spawn may be made. Mr. W. P. AYRES writes lately to tell me that in a great midland garden where the spawn bill used to amount to 18l. or 19l. a year, by saving the spawn as the Parisian growers do, all expense for this article is abolished.

I do not attempt to praise or even duly weigh the merits of the Mushroom—that could only be adequately done by the immortal BRILLAT-SAVARIN. He, however, seems to have somewhat neglected this most precious of *légumes*. None but his serious soul could have approached the subject with the necessary solemnity. Nobody but he who first saw the deep dangers of hurried, thoughtless, and irreverent feeding, could have done justice to its exquisite flavour when in the best condition, or could have explained how deliciously it combined the virtues of herb and flesh, unspeakably superior to either. Let us, in passing, quote one of his aphorisms, contributed to form the *base éternelle à la science*: "*La découverte d'un mets nouveau fait plus pour le bonheur du genre humain que la découverte d'une étoile!*"

Now, I do not hesitate to say that the introduction of the Mushroom into our domestic economy in as great a degree as we have it in our power to produce it, would practically be the addition of a new agent in our *cuisine*, second to none for its delicacy, and unsurpassed for utility. It is true the Mushroom is plentiful in its season, but it is with us, at all seasons when it is not to be gathered in the open air, a luxury to numbers of owners of gardens who have means to grow it. As for the much larger class who ought to be supplied from our markets, they seldom see or taste a Mushroom except when these occur in profusion in our fields, though every cart of stable-manure produced in this great horse-keeping country may, on its way towards decomposition and replenishing the earth, be made a nidus for furnishing many dishes of them.

The illustrations showing the cave-culture of mushrooms are from my "Parks, Promenades, and Gardens of Paris." And the frontispiece is after two large cuts of the mushroom caves of Paris, which appeared in the *Illustrated London News* some time after the appearance of my work. The illustrations of edible fungi are by Mr. WORTHINGTON G. SMITH, who knows and draws these interesting subjects so thoroughly well; and the other figures are by Mr. HODGKIN.

Contents

WHERE MUSHROOMS MAY BE GROWN

THE places in which mushrooms can be grown may be roughly grouped as follows:—1. In the mushroom-house proper. 2. In sheds, cellars, out-houses, stables, railway-arches, &c. 3. In deep caves, like those near Paris, described further on. 4. In the open air, in gardens or fields, on prepared beds. 5. In gardens, among various crops, without any preparation beyond inserting the spawn. 6. In pastures where the mushroom is not already established.

To these I might add another group, illustrated by the case of a Belgian cook who grew a dish of mushrooms in a pair of old wooden shoes; but practically we can treat of nearly every possible mode of growing the mushroom under the above headings.

Chapter I

MUSHROOM CULTURE IN THE MUSHROOM-HOUSE

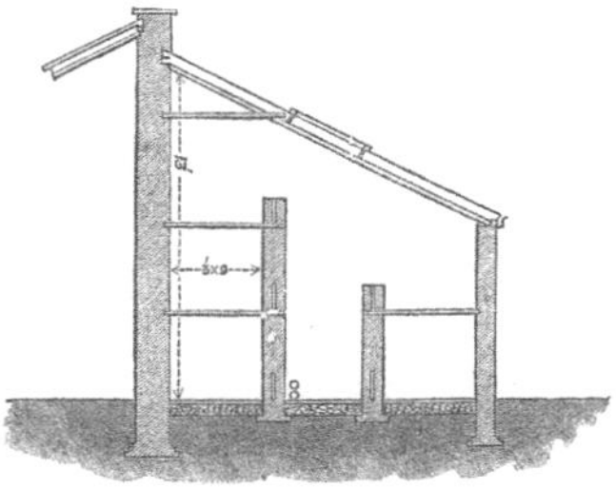

Figure 1 Mushroom-house at back of hothouses

CULTURE in the mushroom-house being the most practised, and, on the whole, the most important phase of the subject, we will first treat of it. And first of the mushroom-house itself. Its construction is very simple: the conditions to be obtained are equable temperature, secured by thick or hollow walls and by a double roof. Figure 1 shows a house designed for me by Mr. Ormson, the well-known horticultural builder.

It is situated at the back of the hothouses, where a flow and return pipe can be run through for artificial heat. The shelves for making the beds upon are of slate 1½ in. thick, or of stone 2½ in. thick, built into the walls, and into brick piers built in cement. Upright slates, to slide in grooves, are placed along the front of the shelves to keep the beds in.

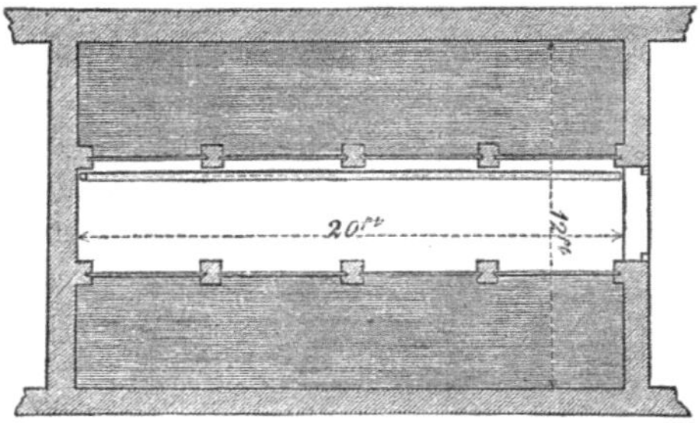

Figure 2 Ground-plan of Preceding

The floor may be of paving tiles, or bricks, laid on concrete: a sky-light or two may be fixed in the roof, for the purpose of admitting a little light, and air when necessary. The engraving (Figure. 2), shows a house of this description, 12 feet wide by 20 feet long, inside measure, but, of course, the length may be extended as circumstances may require.

As it is of importance in mushroom-growing that the air of the house should be kept moderately moist, the underside of a slate or tile roof should be lathed and plastered.

Figure 3 View of unheated mushroom-house.

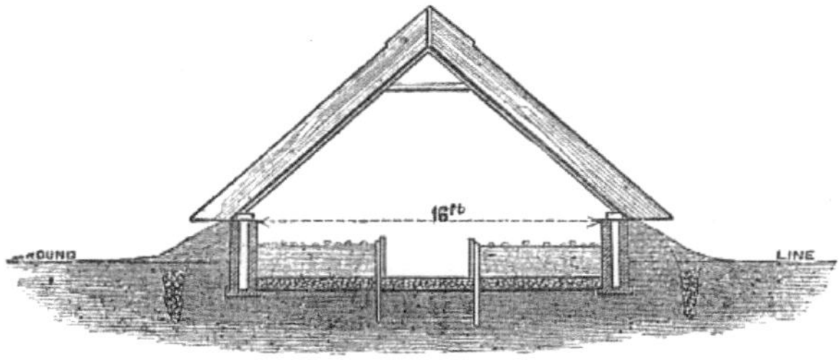

Figure 4 Section of preceding figure.

Figure 3 represents a mushroom-house suitable for people of small means, or those who cannot adopt plan No. 1. It is designed with a view to growing mushrooms during the greater part of the year, without the aid of artificial heat. To this end it is constructed in such a way as not to be affected by changes of the external temperature, as will be seen by the engraving. The walls are hollow, and banked round with the soil excavated from the interior. The roof is thatched with reeds, and the ends stud-work, lined inside with boards, and outside with split larch poles: the cavity to be filled with sawdust or cut straw; a small diamond-shaped ventilator, hung on pivots, to be fixed in each end. The floor may be of concrete, or burnt clay well rammed; and the beds are retained in their place by boards nailed to good oak posts. Care should be taken to put in efficient drains, so that no stagnant damp may exist about the building.

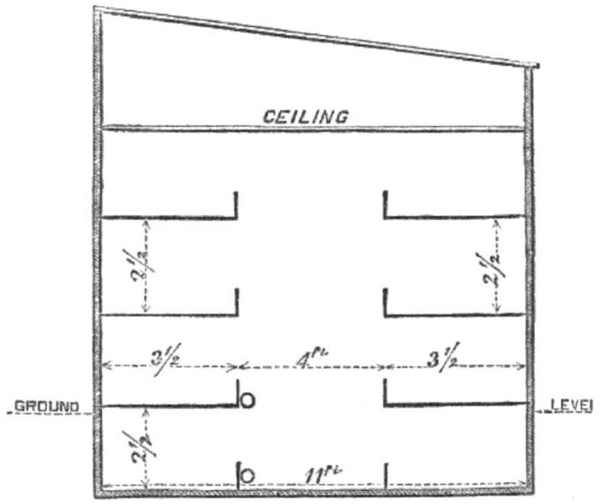

Figure 5 Section of mushroom-house at Frogmore.

Though the preceding cuts show how we may best attain our object, a few more illustrations of mushroom-houses are desirable here. Figures 5 and 6 exhibit the plan of the mushroom-houses at Frogmore, obligingly communicated by Mr. Rose.

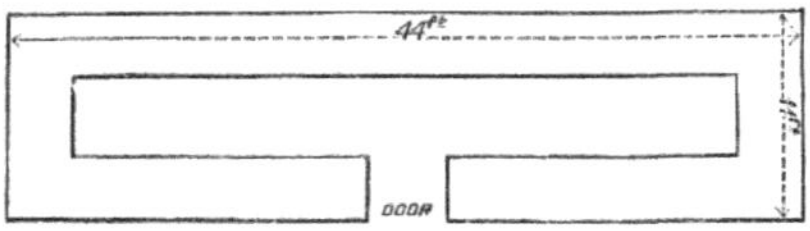

Figure 6 Ground-plan of mushroom-house at Frogmore.

It need hardly be said that in such large mushroom-houses rhubarb and sea-kale may be easily forced, and barbe de capucin, endive, &c. blanched.

A small hot-water apparatus, with a 3-inch flow and return pipe, affords the best means of heating a mushroom-house which is not so situated that it may be heated from the boilers of adjacent hothouses. The best position for the mushroom-house is against a north wall. The usual precautions for guarding against damp walls and floor should be adopted in the case of the mushroom-house, and the walls should be hollow.

Forsyth's mushroom-house is described by the designer in Loudon's *Gardener's Magazine*. Figure 7 is a transverse section, showing the arches under and over the beds, the thoroughfare *a* is the middle, and the position of the hot-water pipes, *c*; *b* is an open shed and general workshop, the receptacle of everything requiring protection, and too clumsy to be otherwise housed.

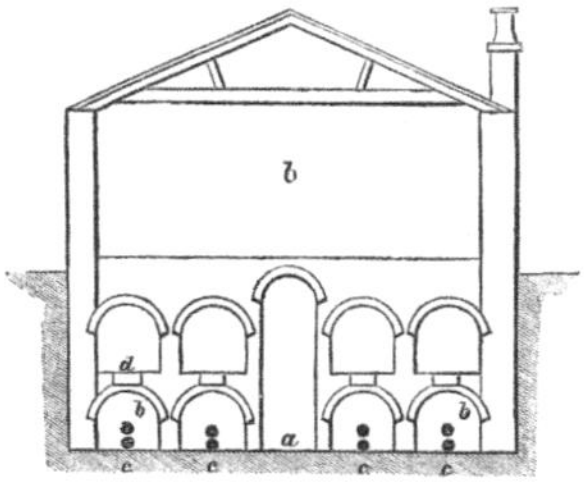

Figure 7 Mushroom-house under shed.

A shed of this description is an indispensable adjunct to every well-ordered garden, and in the present case it serves as a roof to the mushroom-house. In the centre of each vault, shown in Figure 7, a circular ventilator, *d*, 9 in. in diameter, should be made, having a stone and cast-iron stopper, with a folding ring. The whole roof of the mushroom-house is covered over with pavement, which at the same time forms the floor of the shed above. Mr. Forsyth objects to cast-iron shelves "on account of the rust, and to slate shelves, as being cold and damp, and therefore not suitable to the purpose;" but he knows of no objection to shelves built of bricks and mortar, kerbed with hewn stone 3 in. wide, and clamped together with lead.

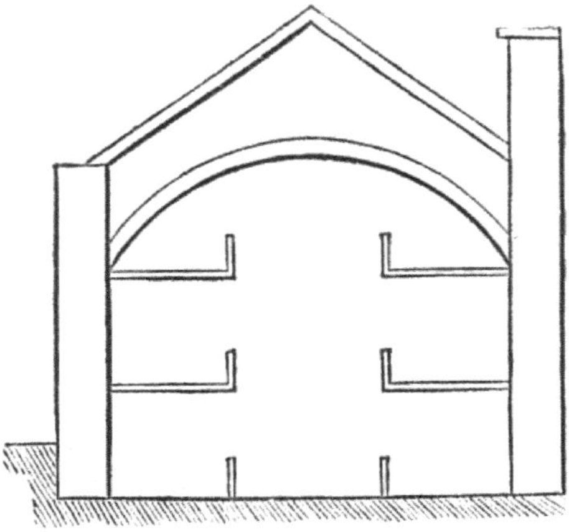

Figure 8 Mushroom-house at Stoke Place.

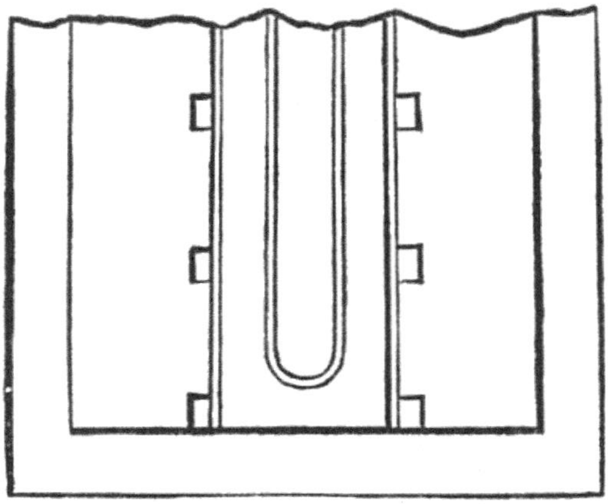

Figure 9

The annexed diagrams (figures. 8 and 9) exhibit the mushroom-houses used at Stoke Place, both for summer and winter use, as described by Macintosh in the "Book of the Garden." "Of course the former is not heated; the latter is, by 4-inch hot-water pipes, which are brought from a boiler constructed to heat at the same time a range of pits for pines, melons, &c., 89 feet long and 7 feet wide. The shelves are close-bottomed to prevent the beds from drying too rapidly, and to require less watering, which Mr. Patrick thinks a very important precaution in mushroom culture. Ventilation is effected by a slide in the door, and a wooden trunk up through the arch and roof, with a slide in it also. We do not exactly see the motive of Mr. Patrick, whom we have long known and esteemed as one of the best gardeners in England, in adopting the span roof over this house, as, from its situation behind the garden wall, a lean-to roof would have been cheaper and carried off the rain-water better. It is rather a novel, but still a good plan, to have the inner roof constructed of a brick arch, as it will of course save the outer one from decay, to which all mushroom-house roofs are liable more than any other kind of garden building. This house struck us at first sight as very complete, excepting in breadth. We should increase it to 9 feet—that is, 3 feet for the breadth of the beds on each side, and the same for the footpath, which at present is inconveniently narrow."

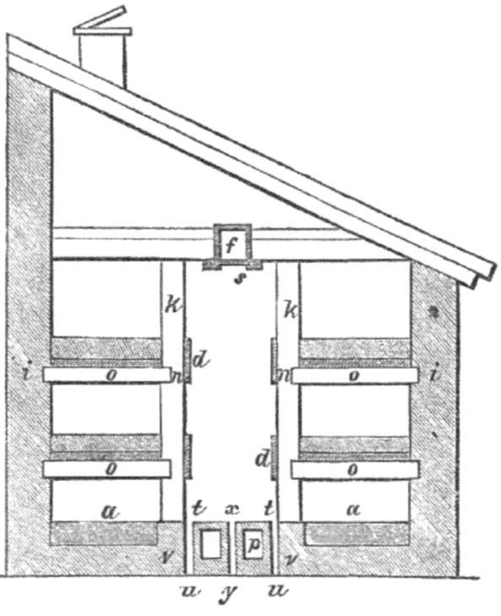

Figure 10 Russian mushroom-house.

The Russian mushroom-house (Figure 10) is thus described by Mr. Oldacre, in the *Horticultural Society's Transactions*, vol. ii. first series. "The outside walls should be 8½ feet high for four heights of beds, and 6½ for three heights, and 10 feet wide inside the walls. This is the most convenient width, as it admits of shelves 3½ feet wide on each side, and affords a space through the middle of the house 3 feet wide, for a double flue and a walk upon it." Hot-water pipes were not in use when this house was erected. "The walls should be 9 inches thick, and the length of the house as may be judged necessary. When the outside of the house is built, place a ceiling over it (as high as the top of the walls) of boards 1 inch thick, and plaster it on the upper side with road sand well wrought together, 1 inch thick, (this will be found superior to lime), leaving square trunks, *f*, in the ceiling 9 inches in width, up the middle of the house, at 6 feet distance from each other, with slides, *s*, under them, to admit and take off air when necessary. This being done, erect two single-brick walls, *v v*, each five bricks high, at the distance of 3½ feet from the outside walls, to hold up the sides of the lower beds, *a a*, and form one side of the air-flue, *t u t u*, leaving 3 feet up the middle, *t x t*, of the house for the floor. Upon these walls, *v v*, lay planks, *t u*, 4½ inches wide and 3 inches thick, in which to mortise the standards, *t k*, which support the shelves. These standards should be 3½ inches square, and placed 4 feet 6 inches asunder, and fastened at the top to the ceiling joists. When the standards are set up, fix the cross-bearers, *i n i n*, that are to support the shelves, *o o*, mortising one end of each into the standards, *n*, the other into the walls, *i*. The first set of bearers should be 2 feet from the floor, and each succeeding set 2 feet from that below it. Having thus fixed the uprights, *t k*, and bearers, *i n*, at such a height as the building will admit, proceed to form the shelves, *o o*, with boards 1½ inches thick, observing to place a board, *d d*, 8 inches broad and 1 inch thick, in the front of each shelf, to support the front of the beds. Fasten this board on the outside standards, that the width of the beds may not be diminished. The shelves being completed, the next thing to be done is the construction of the flue (*p* in section), which should commence at the end of the house next to the door, run parallel to the shelves all the length of the house, and return back to the fireplace, where the chimney should be built; the sides of the flue inside to be of the height of four bricks laid flatways, and 6 inches wide, which will make the width of the flues 15 inches from outside to outside, and leave a cavity, *t u*, on each side betwixt the flue and the walls that are under the shelves, and one, *x y*, up the middle, betwixt the flues, 2 inches wide, to admit the heat into the house from the sides of the flues." The introduction of this form of house by Mr. Oldacre has led to much improvement in our mushroom culture. The first house of this kind erected in England, was built at Shipley, near Derby, in the garden of E. M. Mundy, Esq., by the father of Mr. W. P. Ayres, whose name will be found frequently mentioned in this work. There brick arches were formed for the shelves, and though built more than half a century ago, the house is still in good condition.

Although slate is generally used for the shelves, the adoption of cast-iron gratings for this purpose is well worth a trial, as by this means we may be enabled to cut mushrooms from the under as well as the upper side of the bed.

CHAPTER II

THE PREPARATION OF THE MATERIALS, ETC

BEFORE we deal with the various ways of growing the mushroom, we will speak of the preparation of the material. As stable manure not only furnishes the nutriment, but forms the very soil in which mushrooms are produced artificially, and also supplies the heat which enables us to grow them to perfection at all seasons, by far the most important point connected with their culture is the management of this. It is very simple, but frequently, even by excellent gardeners, considered to require much more trouble and nicety than is really necessary. For example, it is quite common in good gardens to see the droppings collected carefully in some shed, or in the mushroom-house, and turned over almost as tenderly and carefully as the contents of the fruit-room. Good mushrooms are well worth this trouble; but, as it is quite unnecessary, it should not be done except in special cases.

To show the diversity of opinion among excellent mushroom-growers as to the preparation of the manure, I will quote a few of our most trustworthy authorities on the subject. Mr. W. Early, in "How to Grow Mushrooms," lays great stress on the importance of gathering the droppings in a dry state. "Every advantage should be taken of opportunities of securing and placing them in any open shed, or other similar position, where they can be effectually sheltered from rains. In such a place, whilst the process of collecting is going on, every portion should be spread loosely over the floor, in moderate sized ridges, or in any other manner that will allow the air to get amongst it to assist in drying. It should also be tossed over or turned, and lightened up daily for the same purpose, until a sufficiency is gathered together for immediate use."

This may be taken as a sample of the practice very extensively followed in this country. Happily, we have excellent mushroom growers who succeed without all this trouble, as the following remarks of Mr. J. Barnes will show:—"For the last thirty years I have made my beds entirely on the floor in sheds, wheeling in the stable dung as it is brought fresh from the stable, adding a fourth, or a little more than a fourth, of good friable loam, mixing both well together, pressing firmly down, and letting it remain about a week or so untouched. At the end of that time we turn it over, and if we consider it in too strong a state of fermentation we add a little more soil, and then tread down firmly. Very soon the bed is ready to be spawned, and encased in a couple of inches of soil; and in this way we get the finest crops of mushrooms, the beds remaining a long time in bearing. After the beds have been some time, say from six to twelve weeks, in bearing, and begin to get dry, and cease to bear well, we water them thoroughly with very clear liquid manure, made from sheep or deer or cow manure, which seems to start them again into bearing, and then we manage to keep some of the beds in bearing for many months at a time." In the*Field*, Dec. 22, 1868, I stated that the manure for the mushroom-beds in the Royal Gardens, Frogmore, was not prepared in any elaborate way, but simply taken from a great heap fermenting in the yard, any parts of it that had become white from heat being moistened with water, and the whole being mixed with about a fourth part of loam. Mr. Cuthill, an authority on mushroom culture, tells us how the London market gardeners manage with their manure. As the material is brought home from the London stables, the short part is taken out of it, and the long litter is kept for the purpose of covering, as well as for forming the interior of ridges; for all mushroom-beds out of doors are made into ridges. The manure is not allowed to heat before it is put into the beds, if that can be prevented; for previously heated material does not produce such fine mushrooms. The fresher the horse-dung is, the longer the crop will last and every gardener who makes up beds with unheated droppings knows how superior they are to fermented manure.

In his own practice Mr. C. depended a good deal on heavy tramping to "keep down fermentation" when droppings were used in a fresh state. The French, who are great mushroom growers, allow the manure to heat first, but treat it very simply. They prepare it in the open air, first removing any pieces of wood or other extraneous matter that may have been mixed with it, and then place it long and short in beds two feet thick, or a little more, pressing it with the fork. When this is done, the mass or bed is well stamped, then thoroughly watered, and finally again pressed down by stamping. It is left in this state for eight or ten days, by which time it has begun to ferment, after which the bed ought to be well turned over and re-made on the same place, care being taken to place the manure that was near the sides at first towards the centre in the turning and re-making. The mass is now left for another ten days or so, at the end of which time the manure is about in proper condition for making the beds, either in the open air or in the caves. Sometimes it receives three turnings over, especially when the manure is long, and it occupies altogether about six weeks in preparation. As the wide heaps are turned over by the men, a water-cart remains alongside, and any portions of the mass that are dry and white from heat are moistened with water from a rose watering pot. This preparation shortens and mollifies the longer material considerably, mixes the mass well, and it is transferred to the caves in a slightly decomposed, well mixed, and moist, but not wet, condition. The French do not actually hammer or desperately tramp down the beds, as nearly all our writers on mushroom culture recommend, but press it pretty firmly; and I have seen as good crops on their light spongy beds as ever I have on those so firmly tramped down. I might give other striking instances of the diversity of opinion on this subject, but it is needless to multiply them.

My conclusions respecting the preparation of the manure for mushrooms are as follows:—1. That very careful preparation and frequent turning over of the manure undercover are not necessary to success, and that it is quite needless to prepare the manure under cover, except when it is gathered in a very small quantity, so that a heavy rain or snow would saturate it. Where, however, the culture is pursued on a very small scale, and, it may be, only one bed made, it is best to keep it in a covered shed. 2. That carefully picked droppings are not essential, though they may be more convenient. Excellent crops are gathered from beds made with ordinary stable manure, droppings and long materials mixed as they come; but when the manure is used as it comes from the stable, it should be allowed to ferment before being used. 3. That the best way of preparing manure for the general culture of mushrooms indoors, is to gather it in some firm spot, and allow it to lose its fierce heat. As it is usually gathered in an irregular way, precise directions as to turning over cannot well be given; but I am convinced that one turning will suffice when it has arrived at a strong heat, and then it should be thrown together for a week or so, when, in being disturbed and removed to make the bed or beds, its strong heat will be sufficiently subdued. Where large quantities of stable manure are in a fermenting state, there should be little difficulty in selecting material to form a bed at any time. Should it have spent its heat overmuch, it would be easy to revive it with some fresh droppings. 4. That stable manure may be used when fresh, but it should be always mixed with more than a fourth of good loamy soil. If this be kept under cover, or stacked so that it may be had in a rather dry condition, so much the better, especially if the fresh manure, &c., should be over moist. Beds thus made are most suited for cool sheds and the open gardens. 5. That a portion, say nearly one-fifth to one-third, of good and rather dry loam may always be advantageously mixed with the stable manure; the fresher the materials, the more loam should be used. In all cases it helps to solidify the bed, and it is probable that the addition of the loam adds to the fertility and duration of the bed. 6. That a thickness of from one foot to fifteen inches for the beds in an artificially heated house is quite sufficient. Eighteen inches will not be too much for beds made in sheds, though I have seen excellent crops on beds only a foot thick, in common sheds with leaky sides. All beds made indoors should be flat and firmly beaten down, though the absence of firmness is not, as some think, sufficient to account for want of success.

I will now quote a few words from Mr. Ayres on other materials for forming mushroom-beds than stable manure. He has given this, like almost every important subject in the range of horticulture, some attention. First among these may be mentioned sawdust which has been used for bedding horses or for riding-school tracks. Such a substance, thoroughly impregnated with urine and mixed with horse-droppings, forms an excellent material for mushroom-beds, especially if mixed with one-fourth of good fibrous loam. Such materials mixed and fermented together, and thrown into a bed a foot or eighteen inches in thickness, according to the temperature of the shed in which the bed is made, will be found to form capital material for growing this esculent, especially as it retains the heat for a long time. The worst of it is that the material is almost valueless after it has served the first purpose; and used as dung upon light land is rather injurious than otherwise. Then you may use leaves and loam, in the proportion of one part of the latter, in a turfy state, to four or five of fermenting leaves. These may be recently gathered from the trees, and should be allowed to attain a brisk heat before the loam is added, and then, after sweating for a week or ten days, may be turned, mixing the materials intimately together, and then the mass may be formed into a bed. A mushroom-bed of this kind should not be less than fifteen inches in thickness when thoroughly consolidated; and when so managed it will grow mushrooms just as well as dung. The sweepings of our streets and cattle markets, especially those parts that are paved and much frequented by horses—as, for example, cabstands, &c.—if collected when dry, and fermented a little, yield capital material for beds. Here from the cattle market we have the dung of horses, sheep, and cows mixed together in a finely divided state, the heating of which is gentle and regular. Material of this kind procured on dry days, thrown together to ferment once or twice, and then made into well-consolidated beds, will produce mushrooms of the finest quality, and continue in bearing a very long time. It is of the first importance that this material be collected in a dry state, as of course the slush of the streets would not do at all. Equal proportions of street sweepings and fresh leaves, properly fermented and mixed with loam, would perhaps make as good material for growing mushrooms as need be obtained. Of course the sweepings from those parts of the town most frequented by horses will be the best for the purpose I am writing about.

The idea of mushrooms ceasing to be prolific from the exhaustion of the active manure in the bed, I have mooted before. Lately several experiments have been tried which convince me that by taking three portions of recently-gathered leaves to one of turfy loam, and working them well together until the mass attains the desired temperature, sprinkling it, as the work of turning proceeds, with liquid direct from the stables, and forming this into a bed treated in the usual manner, it will give just as good mushrooms as the best horse manure in the world. It is the ammonia that is wanted for this crop, with a gentle heat. Secure these two things, and, with ordinary care, success is certain.

Before making the beds, while the material is in preparation, all particles of old wood, twigs, &c., that are found in the manure should be removed, as indeed should any extraneous matters likely to prove offensive or useless.

The best time for making mushroom-beds, where they are not regularly made in succession throughout the autumn and winter months, as they ought to be where there is abundance of material and a good mushroom-house, is in August and September, as in the early autumn months the natural heat is sufficient to cause the spawn to germinate freely, and beds made then ought to bear freely before and up to Christmas, and during autumn.

When making the bed, the chief object to bear in mind is the equal placing of the material. It should be well mixed and regularly and firmly placed so that the whole may be of a similar texture. Some heavily tramp and pound their beds to secure firmness; moderately done this is beneficial; thoroughly equable pressure with the fork, when the fork can be used, will with the pressure of firm earthing be sufficient; when beds are made on elevated benches in boxes, and in all positions where but a slight body of material is used, and where firmness cannot result from the general pressure of the mass, some kind of pressure with a wooden mallet or the like must be employed.

The beds once made, we next arrive at the spawning, and will first inquire, What is spawn?

CHAPTER III

MUSHROOM-SPAWN

THE first thing we have to determine is, What is spawn? Generally, the spawn, or what in scientific language is called the *mycelium*, is supposed to be analogous to seed, while it really is what may be termed the vegetation of the plant, or something analogous to roots, stems, and leaves of ordinary plants, the visible part or stem, head and gills, of the mushroom being, in fact, the fructification, though in such an apparent preponderance to the other parts. A knowledge of the anatomy and life-history of the mushroom is not necessary to the cultivator, and is not familiar even to those who make of mushrooms a study. We know that the gills are simply surfaces on which germs or spores are produced. The membrane that covers the spore plates of a single mushroom would cover a large space if spread out, and the spores are counted by myriads. We can see them clearly enough under the microscope—can see in what manner they are borne on and fixed to the gills; but of the history of their lives, from the time they fall from the surfaces on which they were born, till the "young mushroom" or inflorescence is vigorously pushing up from the mass of delicate vegetation which they have given rise to in earth or decaying manure, we know nothing. However, the preparation of the spawn, and the subsequent management of it in the mushroom-bed, are the matters which really concern us.

How is spawn obtained in the first instance? It is found in a natural state in half-decomposed manure-heaps, in places where horse-droppings have accumulated and been kept dry, in riding-schools, sheds to which horses have long had access, in "mill tracks" under cover, in pastures, in partially decayed hotbeds, &c., and rarely or never in very moist or saturated materials. This spawn, sometimes termed "natural" in this country, and called by the French "virgin spawn," is the best that can be obtained, and should be used in preference wherever it can be found. To use it, all that has to be done is to divide the material permeated by the white spawn into pieces a few inches square, and say an inch or more thick. They will of course break up irregularly, but all should be used, whether of the size of a bean, or nearly that of the open hand. Then they are inserted into the surface of the mushroom-beds in the ordinary way.

In nearly every country place, and in numerous suburban ones, in fact, in most places where horses are kept, opportunities of finding this spawn occur. Its white, filamentous, and downy threads have the odour of mushrooms, and the spawn is, therefore, very easily recognised. It should be generally known that it need not be used when found, but may be dried, and kept for use in a dry place for years, and has been known to keep as long as fourteen years. It must not be supposed that it is only the hard bricks described further on that keep thus. The French spawn is in much looser and lighter material than that in which we usually find *mycelium* in a natural state, and it keeps quite as long as ours. To preserve spawn found in a natural state, nothing more is required than to take up carefully the parts of the manure in which it is found, not breaking them up more than may be necessary, and placing both large and small pieces loosely in rough shallow hampers. These should be placed in some dry airy loft or shed till thoroughly dry, and afterwards kept in some perfectly dry place, packed in rough boxes till wanted for use.

Figure 11 Brick mushroom-spawn

But inasmuch as in this country, at present, but little mushroom-spawn is required in any one place, the rule is to obtain artificial spawn in the form of hard bricks. This spawn is made from horse-droppings and some cowdung and road scrapings beaten up into a mortar-like consistency in a shed, and then formed into bricks, slightly differing in shape with different makers, but usually thinner and wider than common building bricks. Various recipes are given for mixing the materials for the bricks, and among them the following are about the best:—1. Horse-droppings the chief part, cowdung a fourth, and the remainder loam. 2. Fresh horse-droppings mixed with short litter the greater part, cowdung one third, and the rest mould or loam. 3. Horsedung, cowdung, and loam in equal parts. These bricks are placed in some dry, airy place, and when half dry, a little bit of spawn about as big as a hazel nut, is placed in the centre of each; or sometimes, when the bricks are as wide as long, a particle is put near each corner, just inserted below the surface, and plastered over with the composition of which the bricks are made. When the bricks are nearly dry, they are placed on a hotbed about a foot thick, in a shed or dry place. On this the bricks are piled, or placed rather openly and loosely, and covered over with litter, so that the heat may circulate equably amongst them. The temperature should not rise more than a degree or two above 60 degrees; if it does, it may easily be modified by reducing or removing the covering of litter. The makers frequently examine the bricks during the process, and when the spawn has been found to spread throughout a brick like a fine white mould, it is removed, and allowed to dry for future use in a dark, dry place. If allowed to go further than the fine white mould stage, and form threads and tubercles in the bricks, it has then attained to a higher degree of development than is consistent with preserving its vegetative powers, and therefore it should be removed from the bed in the fine mould stage. This is the kind of mushroom spawn mostly in use in our gardens, and it is usually very hard in texture.

Figure 12 Mill-track mushroom-spawn

There is a kind of spawn used in some gardens called mill-track mushroom-spawn, which is made in a more simple manner than the preceding. It would seem to be simply spawn that has spread through the thoroughly amalgamated droppings of a mill-track. The material is rather soft and free in texture, is usually sold in large and somewhat irregular lumps, and is much used by some cultivators.

Figure 13 Parisian mushroom-spawn

Finally, we have the French mushroom-spawn, which differs from our own in not being in bricks or solid lumps, but in rather light masses of scarcely half decomposed, comparatively loose and dry litter. This spawn is obtained by preparing a little bed as if for mushrooms in the ordinary way, and spawning it with morsels of virgin spawn, if that is obtainable; and then when the spawn has spread through it, the bed is broken up and used for spawning beds in the caves, or dried and preserved for sale. It is sold in small boxes, and is fit for insertion when pulled in rather thin pieces, about half the size of the open hand; but in separating it, it divides into many pieces, of all sizes, every particle of which should be used. The small particles should be strewn broadcast over the bed after the larger pieces have been inserted. This applies to the other kinds. In consequence of the open porous nature of the French mushroom-spawn, it is likely to be immediately affected by the heat and moisture of the genially warm manure forming the mushroom-bed, and on that account alone presents some advantages. It has recently been introduced for the first time, and probably will soon be tested by many growers.

Spawn, in the common sense of the word, may be dispensed with by well amalgamating manure, loam, and old mushroom-beds, or leaf-mould containing traces of spawn, and these formed into beds about a foot thick in the mushroom-house, and covered with earth, produce without any further spawning; but the plan is not so simple or advantageous as that more commonly pursued.

There is no necessity for purchasing artificial spawn at all where mushrooms are regularly grown. Nor is there in any case except at the commencement, or to guard against one's own spawn proving bad. To secure good spawn, we have only to do as the French growers do: take a portion of a bed where it is thoroughly permeated by the spawn and before it begins to bear, and preserve it for future use.

Of the efficacy of this sort of spawn, if any proof were needed in addition to the fine crops the Parisian growers gather, it will be found in the following statement from Mr. Ayres:—

"A short time back, attention was directed to the superior quality of French mushroom-spawn, and as a natural consequence several London seedsmen imported it for sale. Some months back I obtained possession of a stable, and, wishing to grow mushrooms in it, procured a few tons of horse manure, just as it came from the dung-pit of the hotel stables. It was very wet, and consequently when thrown together it heated violently. However, by frequent turning for a week or ten days this tendency was reduced, and then five beds were formed of it, adding one-fourth of perfectly dry soil from a cucumber-house. I say perfectly dry, because the soil had lain in the house for fifteen or eighteen months without receiving a drop of water, and therefore may almost be considered as thoroughly dry. Intimately mixed with the fermenting dung, it had the tendency that I desired—viz., subdued the excessive moisture, and, after the bed had been made up a week, brought it to the temperature necessary to receive the spawn.

"Having great faith in the good qualities of fresh loam from an old pasture for the production of mushrooms of superior quality, I had a quantity dried and warmed. I had a coat of this three inches thick laid over each bed, and then forked carefully in, taking care to mix the soil and dung as intimately as possible. Re-formed and left for a few days the beds attained the necessary warmth; then they were made quite firm, and were ready for spawning.

"For this purpose I had procured two boxes of the French spawn from Messrs. Barr and Sugden, of Covent Garden. It was light, loose, flaky, chaffy stuff, and so dry that I had some fear whether its vegetating power had not been dried out of it. But the spawn had been bought for experiment, and therefore the experiment must be carried out.

"Raking about two inches of the material from the surface of each bed, pieces of the flaky spawn were laid down, at about ten inches or a foot apart, all over the beds; the fine portions of the spawn were then scattered over the beds, patted down firmly with the back of a spade, and then the surface material was returned, and the whole made as firm as possible. In passing, it may not be out of place to remark that spawning in this manner must be guided, or rather governed, by the state of the material of the bed. If it is not sufficiently cooled, it will be safer to make holes in the usual manner for the spawn; but if in a fit state, then I think the broadcast spawning and earthing, as before described, is the best plan. The disturbed portion of the beds having regained its heat, and there being no fear of its *over*heating, the beds were immediately earthed two inches thick with fresh loam, beaten quite firm, and then covered with a thin layer of dry hay.

"Not liking to entrust my chance of mushrooms entirely to the new material, the French spawn, two beds were spawned at the same time and in the same manner with native spawn. Owing to the large size of the stable, and the unusually cold, piercing weather at the end of the year (1869), the beds lost so much heat that I had some misgivings whether they would not prove a failure; but finding, subsequently, that the spawn was working, I gave each bed (the surface being rather dry) a good syringing with water at the temperature of 80 deg., covered it with clean dry mats, and then returned the hay. The beds are now a sheet of the 'pearl of the fields,' some of the patches as large as a cheese-plate, and the whole in most promising condition—so promising that, with proper attention, I have no doubt they will yield a good supply of mushrooms for many months. To secure this continuous bearing, farmyard manure-water and salt, at proper times, should not be spared; while, as soon as the flush of the first crop is over, the beds may receive a thorough soaking of manure-water at a temperature of not less than 80 deg., be re-earthed with fresh soil, and covered down with mats and hay. In this manner we always get a second crop little inferior to the first one, and sometimes much superior."

CHAPTER IV

SPAWNING AND AFTER-TREATMENT

Heat and Protection

THE temperature of the material of the beds should never, at spawning time, exceed 80 degrees Fahr.—about 70 is the most suitable regular temperature; and that of the mushroom-house should range between 50 and 60 degrees—not lower than 50. Assuming the materials to have been turned once after having heated, and again disturbed previous to being made into beds, they ought to be in a condition for spawning from ten to twelve days after being put together. It need hardly be said that this regularity of temperature can only be secured in properly-formed mushroom-houses. Where mushrooms are grown in these, with double ceilings and close-fitting shutters and doors, almost impervious to external influences, and where fresh beds are made from time to time, little or no artificial heat from pipes is required, though it is as well to have some at command in the case of unusually severe weather, or a break in the succession of beds, which would cause a deficiency of heat from fermenting materials. A covering of hay or dry litter is necessary for beds formed in the open air, and also for beds made in cool, half-open sheds; but not for those in regularly heated mushroom-houses or caves, in which there is a still, steady temperature. It should be about a foot thick, and should be immediately removed when it becomes wet or mouldy. This covering should be applied whenever the temperature of the bed begins to fall.

It should not be used in any case where the temperature will permit of dispensing with it, as it is troublesome, and sometimes encourages insects. The heat of a bed may be reduced by opening holes six or eight inches deep with a thick pointed dibber, here and there, but it is only in exceptional cases that this is advisable, and it is desirable to husband all the ammonia and heat of the bed. The earthing over and firming of a bed has a tendency to subdue the heat in it. Where large sloping beds, say three feet deep at back, are made against the wall, I have seen Λ-shaped crates put beneath them at six feet apart, so as to permit of heating them by fresh supplies of manure. It is, however, a plan possessing little claim to general use. It is best not to depend on the hand, as is commonly done, for ascertaining the heat of the beds. Thermometers fixed on sticks of convenient size, to thrust in the beds, are sold, and remove all excuse for vagueness in this matter. Coverings of litter are sometimes useful in "drawing-up the heat" in a bed that has become somewhat chilled.

Spawning

This is the phase of the culture which requires most attention, as to get the spawn to run regularly through the bed is to be nearly certain of securing a good crop. In this respect there do not seem to be so many differences of opinion among mushroom growers. Some, indeed, spawn immediately after the bed is made up; but, except where the materials are such as will not heat to more than 80 degrees, this is uncertain, or in other words bad, practice.

The important thing should be to ascertain if the spawn spreads through the bed properly. The usual practice is to earth up the bed immediately or very soon after it is spawned, and not a few take no further notice of the bed or beds till the time arrives when the mushrooms ought to appear. A better plan is not to finally earth the bed until the spawn is seen beginning to spread its white filaments through the mass; and should it fail to begin to do this in eight or ten days after spawning—the conditions being favourable—it is then better to insert fresh spawn or to re-make the bed, adding fresh materials if it be found to fail from being too cold. If people generally were to see whether the spawn had "taken" freely, instead of waiting for many weeks, not knowing whether it had or not, there would be fewer disappointments in mushroom culture.

The ordinary spawn bricks should be broken into pieces, say from about the size of walnuts to that of eggs; they do not break up into regular portions. Spawn in the more natural form in which we take it from the old beds, and in which it is used by the French, is ready to be inserted into the bed without any further manipulation. I believe this kind of spawn spreads more rapidly through the beds than our own brick spawn, and is, on the whole, much more desirable. As it is usually very dry it is a good plan to place some of it in the mushroom-house a few days before spawning, so that it may begin to absorb moisture. A dark place in a warm house, or gentle hotbed, would do as well, but in no case should it be done more than three days before spawning time. At spawning this might with advantage be mixed with some that has not gone through this process. A bushel of the ordinary brick spawn will suffice to spawn about one hundred square feet. All spawn should be inserted near the surface, just buried in the materials of which the bed is made. The thin flakes of spawn which the French use, and which are usually nearly the length and breadth of the open hand, are generally inserted into the bed edgeways, or in a direction slanting upwards, so that while one edge of the piece is buried three or four inches in the bed, the other is seen peeping through at the surface. Thus each flake of spawn is exposed to a slight difference of temperature, and, being thin and spongy enough to be immediately impregnated with the moist warmth of the beds, takes quickly and well. As to any particular mode of inserting the spawn, little need be said; if the bed be beaten so hard as many recommend, and which I do not believe to be at all necessary, a dibber will be required to insert the spawn; if not, it may be readily inserted with a trowel or with the hand. It is a good plan to use a mixture of two kinds of spawn.

Soil

As regards the kind of soil used in earthing, it is not of nearly so much importance as is generally supposed; almost any soil will do; but those having heaps of good maiden loam laid by for gardening purposes will prefer to use a coating of that. I believe that any ordinary garden soil would do, and feel certain that it is a mistake to bestow the least trouble on procuring any particular kind of soil from a distance. The beds in the caves around Paris are covered over with a white putty-like substance, which would be sufficient to shake the nerves of any British mushroom-grower accustomed to his coatings of mellow loam. It is simply the fine rubbish from the stone breakage moistened, and smoothly and firmly pressed over the beds. We, if shown this on a bed that had failed, would assuredly attribute it to the "stuff" with which the bed was covered, though finer crops than these little beds yield it would be impossible to find. I notice this subject so that failures may be traced to their true causes, and not attributed to matters which really have but slight influence. The final covering of from one to two inches of loam or other soil should not be applied till the spawn has begun to spread through the bed, but a very thin layer of dryish loam may be placed on with advantage just after spawning has taken place, as it will serve to make the surface of a more equable temperature. It is a mistake to suppose that a deep covering is of any advantage. The final earthing should be of soil sufficiently moist or moistened to permit of its being pressed into a firm surface. However, unless it is exceptionally dry, a mere sprinkling of water will suffice.

Watering

As the materials of mushroom-beds are generally moist, and as but little evaporation can take place in the structures in which they are usually grown, water is rarely necessary, and should not be applied until the surface of bed and soil are really dry. It should then be given copiously, enough to well moisten the bed, and it should be soft water heated to a temperature of 80 degrees given with a fine rose, and steadily and patiently applied equably over the whole surface of the bed. Waterings that merely wet the surface and saturate the crevices or lower parts of the bed are of no use. If one drenching is not sufficient to moisten the bed properly, another should be given. The flat form of bed is of course much more easily watered, and is on the whole the best for beds under cover. The position of beds will have a great influence on the quantity of water they require, so that it is almost impossible to give precise directions on this head; but I can scarcely conceive a case in which it will be necessary before six or eight weeks after the formation of a bed, and I have seen fine crops gathered without a single watering having been given. In watering old beds one ounce of guano to the gallon of water will prove beneficial.

Vermin in Mushroom Beds

Woodlice are the greatest pests the mushroom-grower has to dispose of, and the most effective way of getting rid of them is by destroying them with boiling water. The surface of the bed being firm and covered with smooth firm soil, the only likely place to afford these creatures the interstices they usually retire into when disturbed, or when not employed in eating the head of every little mushroom that presents itself, is round the edges of the bed, and in the slit which often occurs between the bed and wall or sides of the shelves that support it. There they are likely to be found in great numbers, and may be destroyed wholesale by pouring boiling water all along the crack. If the beds be covered with hay or litter, it will be necessary to remove this and allow them time to retreat into their hiding places; and if the beds are made in any position that permits of the woodlice hiding in other places than the interstices round them, these places should be sought out, marked, and receive a searching dose of the scalding water all at the same time. It need hardly be added that, as it is not mushrooms, but creatures that rival ourselves in their love of mushrooms, that we wish to annihilate, the scalding water must not in any case be applied to the surface of the bed. If on the surface of old or dry beds, or those from which a good many mushrooms have been cut or pulled, there are any loose hollows or crevices in which the woodlice can take shelter, they should be sought out, cleared of vermin, levelled up, and made firm, so that the enemy cannot take up a position in which we cannot attack him. Should this plan fail, half an ounce of sugar of lead, mixed with a handful of oatmeal and laid in their tracks, will quickly destroy the pests.

The small mite is most destructive in a high temperature, and in summer, Mr. Cuthill says, "the maggot" will not breed in a house where the temperature does not exceed sixty degrees, and it is in hot, dry, and half-neglected houses that this pest is usually seen in summer. At that season there is little need to grow mushrooms indoors, and how they may be produced otherwise in great abundance is explained further on. The entrance of rats should also be guarded against.

Mushroom-beds come into bearing about six weeks from the time of spawning, and remain in bearing from two to five months, according to the position in which they are made, and the attention paid to them.

Treatment of Old Beds

Upon the continuous bearing qualities of a mushroom bed a word may be said. It may savour of the ridiculous to say that a plant growing upon a dung bed may fail from the want of manure. Yet such is literally and positively the fact. Beds become worn out, the produce small and spindly, and we directly do away with them and make fresh ones. Instead of doing this, give the bed a thorough soaking of stable urine and water, at the temperature of 80 degrees, using the urine in the proportion of one part to five of soft water, and adding a wine-glassful of salt to each canful; then coat the bed with fresh sod, cover it down with mats so as to promote the heating, and a second crop as good as the first may be obtained. In this matter I speak from experience, and Mr. Ingram, at Belvoir, has followed the same plan for many years with the most satisfactory result.

Gathering the Crop

Gatherings should frequently take place, especially where the culture is pursued on a large scale. Where there are several beds in bearing, the mushrooms should be gathered every morning. In all cases they should be pulled or twisted out, never cut out, so as to leave decaying stumps in the beds. The holes made by pulling out the mushrooms should be filled with a little fine loam, of which a small heap may be kept in the house for this purpose.

Cleansing the House

A word as to the necessity of a thorough annual cleansing of the mushroom-house. The fact that the French cave-cultivators find it necessary to shift from cave to cave, and find that after a cave has been in use a certain time, mushrooms cease to be produced in it, should act as a caution in this respect. In summer, when there is no need to attempt the culture indoors, the house should be thoroughly cleaned out, lime-whited, every surface scraped and washed, and the house freely opened, so as to thoroughly sweeten it.

CHAPTER V

CULTURE IN SHEDS, CELLARS, ARCHES, OUTHOUSES, AND ALL ENCLOSED STRUCTURES OTHER THAN THE MUSHROOM-HOUSE

MUSHROOMS may be, and are, grown to perfection in many less ambitious structures than the mushroom-house proper. Any species of outhouse will do for the autumn and early winter crops. One of the best crops I have ever seen was grown in a dry and unused coach-house. Mr. Robert Fish grows all his crops in a long, low, rude thatched shed, open in front—the beds flat, in a continuous line against a wall, and enclosed by a low board. Mr. Cuthill, who wrote on mushrooms, and who used to grow them very well, grew his in rude sheds placed against walls. It matters not in the least if the shed be open or ventilated here and there, especially for autumn crops, as I have seen admirable crops in low outhouses searched by every gust, and not heated by flues. The beds in these should always be covered with hay. Mushrooms may be grown in cellars; but cellars being commonly under houses, they are not exactly the places to which people like to convey the materials necessary for the making of mushroom-beds. Where they occur away from a dwelling-house, this objection will not hold good. In some cases it might be obviated by making the beds in rough boxes, say 3½ ft. long by 1½ ft. wide, and afterwards introducing them into the cellar. Railway or other arches, or any dry and empty structures, may be used for mushroom-growing.

"The construction," says Mr. William Ingram, of Belvoir, in a letter to the *Field*, "of efficient mushroom-houses is sufficiently understood by most of our hothouse-builders and by gardeners; but the economical adaptation of places which already exist is a matter which may with the greatest advantage be discussed, as there are hundreds of persons about whose establishments may be found outhouses, cellars, quarries, or sheds, capable of conversion into mushroom-houses, who would be very glad to be taught the method of growing mushrooms, and to have the simple principles that should govern the construction of mushroom-houses explained.

"There are few large farmsteads that are without an unconsidered place which could be readily adapted for the purpose of growing mushrooms; and farmers possess the material at hand, horse manure, which would not suffer great deterioration if employed in first raising a crop of mushrooms. Country brewing establishments have equal conveniences and opportunities. By relating the means by which I have been for several years able to raise large quantities of excellent mushrooms, in a place originally but ill adapted for the purpose, I may induce some of those persons who desire the luxury of what Soyer called 'the Pearl of the Fields,' to turn their attention to the subject of their growth.

"I had a large, open, airy shed at command, but it was liable to be affected by changes in the weather, and was altogether too draughty and cold in winter, and too hot in summer. I built within this shed, with rough fir boards, an inner shed, 18 ft. long, 6 ft. wide, and 8 ft. in height; two receptacles for beds were formed, one on the floor, the other above it: and to give the requisite heat in winter, I passed a flue, formed of 9-in. socket pipes, through the house; with this I can always command an adequate amount of heat. The material of which the beds are formed is chiefly droppings, collected from an enclosed and covered exercise ground. These droppings are trampled by the horses, and mixed with straw broken up with the manure by the passage of the horses.

"When first collected it is piled up in a large heap, in a perfectly dry state, and when wanted for the bed is thrown out, sprinkled with water, and fermented for about a week; while hot, it is taken to the house, and as it is thrown in is mixed with a small quantity of soil of a loamy character, and a barrow-load of leaf soil. It is then pressed into as compact a mass as possible by a rammer or mallet, building it up until it forms a bed 10 in. thick in front and 20 in. at the back. After a bed formed of this description of materials has been thus put together, rapid fermentation takes place; and when the most violent fermentative action has passed, and a temperature of 80° is found in the bed, spawn is put into it by means of a dibber. I employ brick spawn obtained from good makers, but, to vary and possibly prolong the period of production, I introduce a certain quantity of spawn saved from old beds. This is longer in its development than the made spawn, and appears as a subsidiary crop. After the bed is spawned, a covering of compact loamy soil is spread on the surface, 1½ in. to 2 in. in thickness, and well beaten upon it so as to form a smooth and hard crust. A temperature ranging from 50° to 60° should be maintained in the house. A lower temperature abstracts the heat from the bed more rapidly.

"When the mushrooms begin to exhibit weakness, as after the bed has produced a certain quantity they will do, from the exhaustion of the more stimulating portions of the manure, I find it an excellent practice to administer a sprinkling of water in which a handful of salt has been thrown (that quantity of salt to a three-gallon can). Saltpetre, though in much smaller quantities, is equally valuable given in the same way. The practice I have described relates to the winter cultivation of mushrooms."

Many instances of perfect success like the preceding could be quoted. Here is one from Mr. W. P. Ayres:—

"You will be glad to hear that we have on the outskirts of this town (Nottingham) a grower of mushrooms (Mr. Cookson, Mansfield Road) who vies with the French growers, especially if the means of growth be taken into consideration. The place he occupies was formerly the pleasure garden of a large hotel, where the proprietor would occasionally, in the summer season, treat his friends and patrons to an *al fresco* entertainment. For this purpose a range of summer-houses was built, consisting of brick arches, say 12 feet deep, 6 feet wide, and a little more in height. Close adjoining is a small sandstone-rock cellar, which used to serve for drinkables in the summer and potatoes in the winter.

"Some twelve months ago these premises and the house adjoining fell into the occupation of a gardener, who, though he had a licence to the house, fancied he might turn the arches to a better purpose, and hence he devoted them to mushroom beds. As it was necessary that the arches should be closed, a wall about three feet high was built in the rudest manner parallel with their front, but six feet from it, and from that a roof of rough timber was thrown, and covered with asphalted felt. Here, however, was a mistake; for, the building standing due south, when the sun fell upon it the atmosphere became rather 'tarry'—so much so that the mushrooms refused to grow in it. That wore off after a time, and from a bed not more than thirty yards square the tenant told me he had cut more than 25*l.* worth of mushrooms. When I saw the beds they might be considered spent, the flush of early youth was over; but still the crop was most wonderful, especially considering the means at command.

"In the rock cellar the small beds were a pavement of splendid mushrooms, many of them as large over as a cheese-plate, and thick in proportion. In the garden is a barn—four walls with a roof over them, the latter so rude that it was only in fair weather that it could be called waterproof. In this place which may be 25 ft. long by 15 ft. wide, two tiers of beds have been put up, the roof has been made waterproof, a common brick flue put through it, and, at the time I saw them, more promising beds could not be desired. Here again, you will perceive expensive appliances are not necessary for the production of mushrooms."

Stables and like structures offer capital positions in which successful mushroom culture may be carried out with ease.

If it is possible, and we know it is not only possible but easy, to grow mushrooms in boxes a few feet long and a foot or eighteen inches wide, and the same depth, it is clear that there can be no difficulty about growing them in abundance in such a manner as that shown in the accompanying engraving. This mode was actually practised with great success by the Baron Joseph d'Hoogvorst, of Limmel.

Figure 14 Mushroom culture on shelves in stable.

The culture was carried out in neatly fitted-up wooden boxes, so arranged that they might be shrouded with canvas curtains as shown in the engraving, so that at first sight one would not suppose that mushroom culture was carried on there. No evil results as regards the creation of an unhealthy atmosphere accompanied the attempt. The beds were formed much in the usual way from the droppings of highly fed horses. Now there can be no doubt that a similar mode of growing mushrooms could be carried out in the stables or some adjacent building in hundreds of places apart from the garden and the gardener altogether. Given the materials and some position, however contracted, in which to carry out the culture, and both these things are surely to be had almost in every place where there is a stable, the rest is so simple that any stableman or boy could carry it out. We know that these individuals, as a class, are not much given to botanical or horticultural studies, but no doubt the prospect of an occasional half-dozen fresh mushrooms on the gridiron would give them most praiseworthy interest in the culture. The only objection to it is, or might be, that once they were at home in the culture, the gardener would be very likely to fall short of materials for his hotbeds. An empty loft, or any other covered structure could be employed as well as the stable or an empty coach-house. Apart altogether from utilizing the walls of the stable, as the Baron did, empty stalls frequently present an opportunity of growing mushrooms in quantity. These remarks apply to stables in cities and towns, as well as in the country; indeed in cities, particularly in London, stable manure is usually so plentiful that it is much easier to obtain and much cheaper than in the country, so that even those in London having suitable places for growing mushrooms, but not keeping horses regularly or at all, could have no difficulty in procuring abundance of materials.

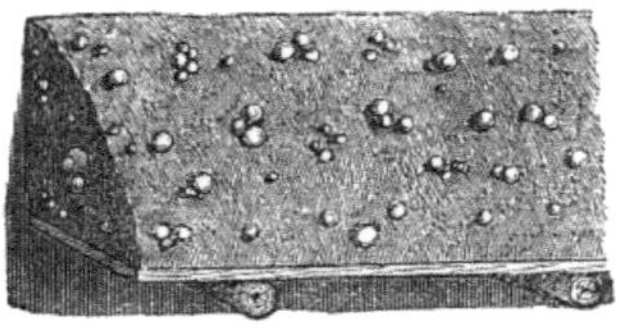

Figure 15 Mushroom-bed on rude shelf against wall of cellar.

The French often cultivate mushrooms in cellars as well as in the caves described in the next chapter. Preference should be given to a dry warm cellar; it should be as dark as possible, and exposed to no draughts. Beds can be made in cellars in many ways. Those made in the middle should always be formed with two sides, while those against the walls should only be half as thick, on account of their having only one useful side. It is also possible to arrange them on shelves, one above the other. For this purpose strong bars of iron are driven into the walls, upon which are placed shelves of the proper size covered with earth, upon which is formed a bed, that is treated exactly as those made upon the ground. These beds are just as productive as any of the other kinds. They may even be made on the bottoms of casks, which should be at least two feet six in diameter; and they are built up in the shape of a sugarloaf, about three feet in height, and the pieces of spawn are placed an inch and a quarter deep, and sixteen inches apart. A barrel is sawn crossways into two pieces, each forming a tub. Holes are made in the bottom of each, and a thin layer of good soil is spread over them inside. They are then filled with good well-prepared stable manure, just like that used in the case of ordinary mushroom-beds, the different layers of dung in each tub being well pressed down. When the tub is half full, six or seven good pieces of spawn are placed on the surface, and the remainder is piled up with manure, which is well pressed down, the operation being completed by giving to the heap the form of a dome. The tubs thus prepared are placed in a perfectly dark part of a cellar, and eight or ten days afterwards the dung is taken up until the spawn is visible, in order to see whether it has commenced to vegetate and develop little filaments. If the spawn has spread, the surface must be covered with soil, care being taken to use only that which is fresh and properly prepared. In this or any like way there should be no difficulty in growing mushrooms: the boxes or tubs could be filled anywhere, and then carried into the spare cellars, &c. In this way objections against steaming manure might in many cases be got over.

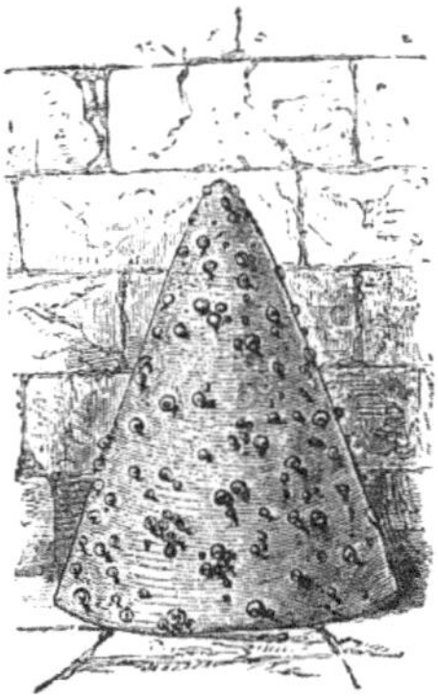

Figure 16 Pyramidal mushroom-bed on floor of cellar.

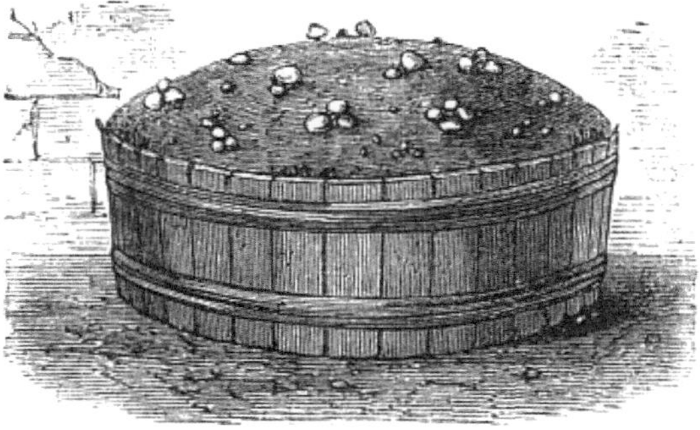

Figure 17 Mushrooms grown in bottom of old cask.

Among the many and various structures in which mushrooms may be grown, but which we rarely see utilized for that purpose, may be mentioned all kinds of greenhouses, stoves, pits, and frames. Some of the best crops I have ever seen were in cold greenhouses almost too ruinous to grow anything else. In mid-winter the floors of all houses in which a genial temperature is kept up for forcing or other purposes, offer excellent positions for producing mushrooms quickly and abundantly. Small ridge-like beds might be made on the floor of these, and, with the genial temperature usually kept up in such places, would probably come into bearing a month or so after being spawned. How often, for example, do we notice the floors of large vineries, in mid-winter or very early spring, quite bare, especially after the vines are started.

Now just at that season the genial heat that would be given off from the slightly fermenting materials used for the mushroom-bed is that which would be most congenial to the tender breaking vines, and with a little attention in this way a first-rate crop of mushrooms could always be gathered from the early vinery, and in houses where no artificial heat was applied they could also be grown abundantly. A covering of hay would, however, be necessary in cold houses in mid-winter, to prevent excessive variation of the temperature, and also in spring and summer to prevent excessive drying or scorching of the beds by a hot sun. I have even seen excellent crops grown on the floor in an old lean-to house, the beds covered with a foot or so of hay, occasionally sprinkled with water to prevent excessive heat on the surface of the bed. In small places where every foot of space in the glass-house is likely to be occupied with plants, it is not easy to carry out the foregoing suggestions, but even if a small early vinery were occupied with plants, it would be desirable and practicable to introduce a series of rough boxes devoted to mushroom culture.

Apart from empty greenhouses altogether, the space beneath the stages in numbers of glass-houses of every type may be utilized for the production of mushrooms. These positions are usually unoccupied, occasionally they are used for storing fuchsias, &c. in winter, but very seldom are they turned to so good account as they might be in the way I recommend. The stage in the small greenhouse is frequently elevated so that there is plenty of room to get beneath it: if at the back or end there is no way of walking readily under the stage, an opening should be made. The only difficulty that could possibly occur would arise from the drip from the plants on the stage above. This, however, can be easily guarded against by spreading a piece of tarpaulin or oil-canvas over the bed or beds. With beds properly made, a coat of dry hay or litter, and a piece of tarpaulin, every owner of anything in the shape of a greenhouse with a stage in it may grow mushrooms throughout the autumn, winter, and spring months, and even in summer by keeping the surface of the hay or litter moist. Of course, if there be room for but one bed, a succession cannot be kept up, and in this case a bed should be made in autumn, which, if well managed, should be in full bearing for a month or six weeks before and after Christmas. There are, however, numerous spaces such as those alluded to where there is room to make a succession of beds. No person having but one greenhouse need fear much or any inconvenience from the odour of the manure—at least, not after the beds are earthed. The couple of inches of soil over the manure would absorb any vapour given off by the bed.

Wherever the cultivation of cucumbers or melons in pits or frames is carried out, nothing can be easier than to grow large crops of mushrooms after the melons, &c. are cleared away. The spawn may be inserted over the surface of the little mounds usually made for the reception of the young melon plants, and also over the remaining surface of the beds which are generally covered with a few inches of earth. After the melons have done bearing and the haulm is cleared away, the spawn will usually be found to have spread through the deep mass of earth in the beds. As little or no water is given or required while the melons are ripening, a good soaking of tepid water will generally be necessary to encourage the mushrooms to start into profuse bearing. If the season and situation be mild and warm, the lights may be taken off; and if the sun be very strong, the beds may be shaded with canvas or mats. If the season be late and cold it will, on the other hand, be desirable to keep on the lights, and even to cover them in cold weather.

CHAPTER VI

THE CAVE CULTURE OF MUSHROOMS, NEAR PARIS

The most extensive and successful culture of mushrooms in existence is carried on in widely-ramifying caves far beneath the surface in the vicinity of Paris. To give the reader as good an idea of it as I can we must visit one of the great "Mushroom caves" at Montrouge, just outside the fortifications of Paris, on the southern side. The surface of the ground is mostly cropped with wheat; but here and there lie, ready to be transported to Paris, blocks of white stone, which have recently been brought to the surface through coalpit-like openings. There is nothing like a "quarry," as we understand it, to be seen; the stone is extracted as we extract coal, and with no interference whatever with the surface of the ground. We find a "champignonniste" after some trouble, and he accompanies us across some fields to the entrance of his subterranean garden. It is a circular opening like the mouth of an old well, but from it protrudes the head of a thick pole with sticks thrust through it. This pole, the base of which rests in darkness sixty feet below, is the easiest and indeed the only way by which human beings can get into the mine. I had an idea that one might enter sideways and in a more agreeable manner, but it was not so. Down the shaky pole my guide creeps, I follow, and soon reach the bottom, from which little passages radiate. A few little lamps fixed on pointed sticks are placed below, and, arming ourselves with one each, we slowly commence exploring dark, still, tortuous passages. I have heard that the first individual who commenced mushroom-growing in these catacomb-like burrowings was one who, at a particularly glorious epoch of the history of France, when a great many more brave garçons went to fight than returned from the victory, preferred, strange to say, to stay at home and hide himself rather than form a unit in "battle's magnificently stern array." Industrious and discreet youth! You deserve being held up as an example as much as the busy bee that improves each "shining hour."

Figure 18 Mushroom-cave, 70 feet beneath the surface, at Montrouge, near Paris, July, 1868.

The passages are narrow, and occasionally we have to stoop. On each hand there are little narrow beds of half-decomposed stable manure running along the wall. These have been made quite recently, and have not yet been spawned. Presently we arrive at others in which the spawn has been placed, and is "taking" freely. The spawn in these caves is introduced into the little beds in flakes taken from an old bed, or, still better, from a heap of stable manure in which it occurs naturally. Such spawn is preferred, and considered much more valuable than that taken from old beds. Of spawn in the form of bricks, such as is used in England, there is none.

Figure 19 Newly-made bed against wall of cave.

The champignonniste pointed with pride to the way in which the flakes of spawn had begun to spread through the little beds, and passed on—sometimes stooping very low to avoid the pointed stones in the roof—to where the beds were in a more advanced state. Here we saw little, smooth, putty-coloured ridges running along the sides of the passages, and wherever the rocky subway became as large as a small bedroom two or three little beds were placed parallel to each other. These beds were new, and dotted all over with mushrooms no bigger than sweet pea seeds, affording an excellent prospect of a crop. Each bed contains a much smaller body of manure than is ever the case in our gardens. They are not more than twenty inches high, and about the same width at the base; while those against the sides of the passages are not so large as those placed in the open spaces. The soil, with which they are covered to the depth of about an inch, is nearly white, and is simply sifted from the rubbish of the stone-cutters above, giving the recently-made bed the appearance of being covered with putty.

Although we are from seventy to eighty feet below the surface of the ground, everything looks quite neat—in fact, very much more so than could have been expected, not a particle of litter being met with. A certain length of bed is made every day in the year, and as the men finish one gallery or series of galleries at a time, the beds in each have a similar character. As we proceed to those in full bearing, creeping up and down narrow passages, winding always between the two little narrow beds against the wall on each side, and passing now and then through wider nooks filled with two or three little beds, daylight is again seen. This time it comes through another well-like shaft, formerly used for getting up the stone, but now for throwing down the requisite materials into the cave. At the bottom lies a large heap of the white earth before alluded to, and a barrel of water—for gentle waterings are required in the quiet, cool, black stillness of these caves, as well as in mushroom-houses on the upper crust.

Once more we plunge into a passage as dark as ink, and find ourselves between two lines of beds in full bearing, the beautiful white button-like mushrooms appearing everywhere in profusion along the sides of the diminutive beds, something like the drills which farmers make for green crops. As the proprietor goes along he removes sundry bunches that are in perfection, and leaves them on the spot, so that they may be collected with the rest for to-morrow's market. He gathers largely every day, occasionally sending more than 400 lb. weight per day, the average being about 300 lb.

Figure 20 View in mushroom-cave.

A moment more and we are in an open space, a sort of chamber, say 20 feet by 12, and here the little beds are arranged in parallel lines, an alley of not more than four inches separating them, the sides of the beds being literally blistered all over with mushrooms. There is one exception; on half of the bed and for about ten feet along, the little mushrooms have appeared and are appearing, but they never get larger than a pea, and shrivel away, "bewitched" as it were. At least such was the inference drawn from the cultivator's expression about it. He gravely attributed it to a ridiculously superstitious cause. Frequently the mushrooms grow in bunches or "rocks," as they are called, and in such cases those that compose the little mass are lifted all together.

The sides of one bed here had been almost stripped by the taking away of such bunches, and it is worthy of note that they are not only taken out, root and all, when being gathered, but the very spot in which they grew is scraped out, so as to get rid of every trace of the old bunch, and the space is covered with a little earth from the bottom of the heap. It is the habit to do this in every case, and when the gatherer leaves a small hole from which he has pulled even a solitary mushroom, he fills it with some of the white earth from the base, no doubt intending to gather other mushrooms from the same spots before many weeks are over. The "buttons" look very white, and are apparently of prime quality. The absence of all littery coverings and dust, and the daily gatherings, secure them in what we may term perfect condition. I visited this cave on the 6th of July, 1868, and doubt very much if at that season a more remarkable crop of mushrooms could be anywhere found than was presented in this subterranean chamber—a mere speck in the space devoted to mushroom culture by one individual.

When I state that there are six or seven miles run of mushroom-beds in the ramifications of this cave, and that the owner is but one of a large class who devote themselves to mushroom culture, the reader will have some opportunity of judging of the extent to which it is carried on about Paris. These caves not only supply the wants of the city above them, but those of England and other countries also, large quantities of preserved mushrooms being exported, one house alone sending to our own country no less than 14,000 boxes annually. There were some traces of the teeth of rats on the produce, and it need not be said that these enemies are not agreeable in such a place; but they did not seem to have committed any serious ravages, and are probably only casual visitors, who take the first opportunity of obtaining more varied food than is afforded them by these caves. To traverse the passages any further is needless—there is nothing to be seen but a repetition of the culture above described, every available inch of the cave being occupied. We again find our way to the bottom of the shaft, carefully mount the rather shaky pole one at a time, and again stand in the hot sun in the midst of the ripe wheat.

In traversing the fields two things relating to mushroom culture are to be observed—heaps of white gritty earth, sifted from the *débris* of the white stone, and large heaps of stable manure accumulated for mushroom growing, and undergoing preparation for it. That preparation is different from what we are accustomed to give it. It is ordinary stable manure, or very short stuff, not droppings, and is thrown into heaps four or five feet high, and perhaps thirty feet wide. The men were employed turning this over, the mass being afterwards stamped down with their feet, a water-cart and pots being used to thoroughly water the manure where it is dry and whitish.

As many will feel an interest in the cave culture of the mushroom, and perhaps wish to see it for themselves, I may state that it is difficult to obtain permission to visit the caves, and many persons would not like the look of the "ladder" which affords an entrance. Even with a well-known Parisian horticulturist I had some difficulty in entering them. I was informed that one champignonniste in the same neighbourhood demands the exorbitant price of twenty francs for a visit to his cave. As the visit is the work of some little time, no visitor should put the cultivators to this trouble without offering some slight recompense—say not less than five francs. The above cave is but a sample of many in the immediate neighbourhood of Paris.

We will next visit a mushroom-cave of another type at some little distance from that city. It is situated near Frépillon, Méry-sur-Oise—a place which may be reached in an hour or so by the Chemin de fer du Nord, passing by Enghien, the valley of Montmorency and Pontoise, and alighting at Auvers. There are vast quarries in the neighbourhood, both for building-stone and the plaster so largely used in Paris. The materials are not quarried in the ordinary way by opening up the ground, nor by the method employed at Montrouge and elsewhere in the suburbs of Paris, but so that the interior of the earth looks like a vast gloomy cathedral. In 1867 the mushroom culture was in full force at Méry, and as many as 3000 lbs. a day were sometimes sent from thence to the Paris market; but the mushroom is a thing of peculiar taste, and these quarries are now empty—cleaned out and left to rest. After a time the great quarries seem to become tired of their occupants, or the mushrooms dislike the air; the quarries are then well cleaned out, the very soil where the beds rested being scraped away, and the place left to recruit itself for a year or two. In 1867 M. Renaudot had the extraordinary length of over twenty-one miles of mushroom-beds in one great cave at Méry; last year there were sixteen miles in a cave at Frépillon. This is a clean, lonely village, just touching on the gigantic cemetery which M. Haussmann projected.

Figure 21 Entrance to large subterranean quarry.

The distant view of the entrance to the quarries has much the appearance of an English chalk-pit. But there is a great rude arch cut into the rock, and into this we enter, meeting presently a waggon coming forth with a load of stones, the waggoner with lamp in hand. To the visitor who has seen the mushroom caves near Paris, where it is sometimes necessary to stoop very low to avoid knocking one's head against the roof rocks, the surprise is great on getting a little way in. At least it is so soon as one can see; the darkness is so profound that a few candles or lamps merely make it more visible. The tunnel we traverse is nearly regularly arched, masonry being used here and there, so as to render the support secure and somewhat symmetrical, the arches being flat at the top for six feet or so, and about twenty-five feet high; sometimes five feet higher.

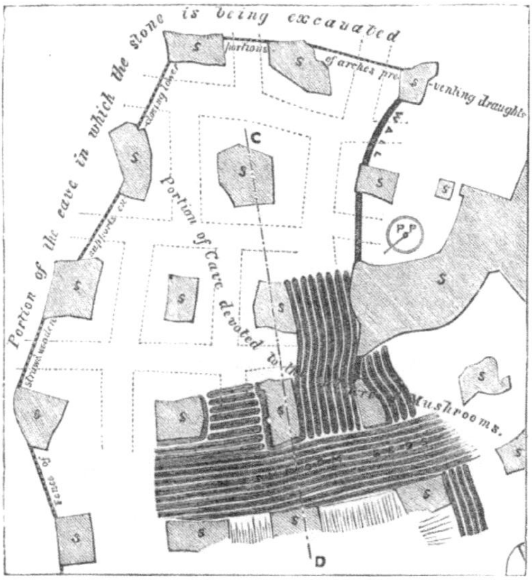

Figure 22 Plan of large subterranean quarry at Fortes Terres, Frépillon. *S*, *S*, *S*, represent the plan of the bases of the huge supporting pillars, and the dotted lines their union with the roof. *D*, *C*, shows the line of the section shown in the following cut, and *P*, place for preparing the plaster. Sept. 1868.

Figure 23 Section following the line *C*, *D*, in Fig. 22.

Presently we turn to the right, and a scene like a vast subterranean rock temple presents itself. At one end are several of us with lamps, admiring the young mushrooms budding all over the rows of beds, which, serpent-like, are long and slim, and coil away into the darkness. At about 150 feet distance there is a group of three men and a boy, each with a lamp, again dispelling the darkness from the mushroom beds, and occupied in placing small quantities of a sort of white clayey sand in the spots whence gatherings have been made a few hours previously. From both sides of this gloomy avenue the dark openings of others depart at short intervals, and the floor of all is covered with mushroom-beds, sometimes running along the passages, sometimes across them. These beds are about twenty-two inches high and as much in diameter, and are covered with silver sand and a sort of white putty-like clay in about equal proportions. Running along in parallel lines, and disappearing from view in the darkness, one knows not what to compare them to, unless it be to barked pine trees in the hold of a ship.

Everywhere on the surface of these little beds small mushrooms were peering forth in quantity; as the beds are regularly gathered from every day, no very large ones are seen. They are preferred when about the size of a chestnut, and are removed root and branch, a small portion of finely sifted earth being placed in each hole, so as to level the bed as in the caves at Montrouge. If the old superstition that a mushroom never grows after being seen by human eyes were true, the trade of a champignonniste would never answer here, as the little budding individuals come within view every day during the gathering and earthing operations. The most perfect cleanliness is observed everywhere in the neighbourhood of these beds, and the whole surface of each avenue is covered by them, leaving passages of ten inches or a foot between the beds. At the time of my visit (Sept. 29, 1868) the crops of the cultivator were reduced to their lowest ebb, and yet about 400 lbs. per day were sent to market. The average daily quantity from this cave is about 880 lbs., and sometimes that is nearly doubled.

In some parts of the cave the work of ripping out the stone by powder and simple machinery continually goes on. The arches follow the veining of the stone, so to speak; their lower parts are of hard stone, the upper ones of soft, except the very top, which is again hard. There is but a slight crust of stone above the apex of each arch, and above that the earth and trees.

It may be supposed that the profits from such an extensive culture are great; and so they are, but the expense is great also. The proprietor informed me that culture on a more limited scale than he pursued last year at Méry gave the best return in proportion to expense, the care and supervision required by so many miles of beds being too great.

Figure 24 Extracting the stone in subterranean quarries

All the manure employed is brought from Paris by rail, as the place is twenty-five miles from that city by road. In the first place, so much per month is paid in Paris for the manure of each horse; then it has to be carted to the railway station and loaded in the waggons; next it is brought to the station of Auvers, and afterwards carted a couple of miles to the quarries, paying a toll for a bridge over the Oise on the way. That surely is difficulty enough for a cultivator to begin with! Then it is placed in great flat heaps a yard deep by about thirty long and ten wide, not far removed from the mouth of the cave, and here it is prepared, turned over and well mixed three times, and as a rule watered twice. About five or six weeks are occupied in the preparation, long manure requiring more time than short. The watering is not usually done regularly over the mass, but chiefly where it is dry and overheated. Every day manure is brought from Paris; every day new beds are made and old ones cleared out—the spent manure being used for garden purposes, particularly in surfacing or mulching, so as to prevent over-radiation from the ground in summer. The chief advantage the cultivator here has is the facility of taking his manure or anything else in or out in carts, as easily as if the beds were made in the open air. Near Paris, on the contrary, everything has to be sent up and down through shafts like those of an old well, and the men have to creep up and down a rough pole like mice. Many men are employed in the culture, the daily examination of sixteen miles of beds being a considerable item in itself. Here and there a barrier in the form of straw nailed between laths may be seen blocking up the great arch to a height of six feet or so. This is to prevent currents of air wandering about through the vast passages.

The mode of preparing the spawn here is entirely different to ours. They prefer virgin spawn—that is to say, spawn found naturally in a heap of manure. But as this material cannot be obtained in sufficient quantity to meet the wants of such extensive growers, they put a small portion of it into a mushroom-bed to spread, and instead of allowing this bed to produce mushrooms, it is all used as spawn, and is valued more than any other. Of course abundance of spawn occurs in the old beds, but it is never used directly. It is, however, frequently employed to spawn a small bed when virgin spawn cannot be obtained. In this case the small bed devoted to the propagation of spawn is placed in the open air, and covered with straw, and as soon as it is permeated with the spawn it is carried into the caves and used. As the making and spawning of beds is a process continually going on, a bed of this sort must be ready at all times. It is never made into bricks as with us, but simply spread through short, partly-decomposed, manure.[A]

I was informed that coal-mines are not adapted for growing mushrooms, and the smallest particle of iron in the beds of manure is avoided by the spawn, a circle around it remaining inert. It is said to be the same with coal. If an evil-disposed workman wishes to injure his employer, he has only to slip along by the beds with a pocketful of rusty old nails, and insert one here and there.

Figure 25 View in old subterranean quarries devoted to mushroom culture, and in the occupation of M. Renaudot. Sept. 29, 1868.

The beds remain in good bearing generally about two months, but sometimes last twice and three times as long. A useful contrivance for facilitating the watering of the beds has lately been invented; it consists of a portable water-cistern to be strapped to the back and fitted with a rose and tubing, so that a workman may carry a larger quantity of water, and apply it more regularly and gently than with

A Mr. Speed, superintendent of the gardens at Chatsworth, has recently prepared his own spawn, as described on p. 73, and with perfect success.

the old-fashioned watering-pots—while one hand is left free to carry the lamp. An iron frame has also been invented, in which the bed is first compressed and shaped, the frame being then reversed and the bed placed in position. Another invention for earthing the beds over as soon as the spawn has taken will soon be in operation if not already so. As on an average 2500 yards of beds are made every month, simple mechanical contrivances to facilitate the operation will prove of the greatest advantage to the cultivator.

In addition to the caves in the localities above alluded to there are other places near Paris where the culture is carried on—notably at Moulin de la Roche, Sous Bicêtre, near St. Germaine, and also at Bagneux. The equability of temperature in the caves renders the culture of the mushroom possible at all seasons; but the best crops are gathered in winter, and consequently that is the best time to see them. I, however, saw abundant crops in the hottest part of the very hot season of 1868. These mushroom caves are under Government supervision, and are regularly inspected like any other mines in which work is going on. As regards the depth at which this culture is practised, it usually varies from twenty to one hundred feet, sometimes reaching one hundred and fifty and one hundred and sixty feet from the surface of the earth. They are so large that sometimes people are lost in them. In one instance the proprietor of a large cave went astray, and it was three days before he was discovered, although soldiers and volunteers in abundance were sent down. Is it possible that in a great mining and excavating country like ours we cannot establish the same kind of industry?

CHAPTER VII

CULTURE ON PREPARED BEDS IN THE OPEN AIR IN GARDENS AND FIELDS

MUSHROOMS may be grown with ease in the open air in gardens; and this is a phase of the culture with which gardeners are not by any means sufficiently conversant. In fact, mushroom-culture in the open air in private gardens may be said not to exist at present, so very rarely is it seen.

In a little pamphlet on mushroom-growing that has lately appeared I find it stated that mushrooms may be grown out of doors "in summer," but nothing about them being grown in the open air in winter. The Paris growers never attempt their culture in summer: the London ones very rarely. It is in winter that their cultivation is carried on in full vigour in the open air. Abundant crops are grown in the open air by the market-gardeners of London and Paris. From their beds mushrooms are gathered in quantities in mid-winter as well as in autumn. The Paris market-gardener does not attempt the culture in mid-summer, and does not think it practicable; but in the hot summer of 1868, and in the midst of the heats of July, I found about half an acre of ground at Brompton covered with mushroom-beds bearing well.

The following illustration is from a sketch taken in Nov. 1869, in market-garden fields, between Kensington and Brompton. The beds, about three and a half feet high and the same in width at the base, are covered with the long straw or litter taken from the stable manure.

Over that is placed old bast mats, or any like materials, to keep the litter in its place, and throw off the rain; the mats being kept in place by tiles, bricks, old boards, or any like objects that may be at hand. This is well shown in my illustration.

Figure 26 Mushroom-beds in market-gardens at Earl's Court, Kensington. November, 1869.

The manure employed is that brought from the London stables, the longer litter being shaken out and put on one side to cover the beds. No care whatever is taken in the preparation of the manure; it is usually made into beds soon after it is brought home and before it is allowed to heat, and then the beds are made in the form of potato-pits and beaten very firm. The beds are spawned when at about a temperature of eighty degrees, the pieces of spawn being placed about a foot or so apart, and it is then immediately earthed, the ordinary soil being used, and the bed covered to a thickness of a couple of inches. The success attained by the market-gardeners of both London and Paris, with the ordinary soil of the place in which the beds may be made, well proves the absurdity of seeking for any particular kind of soil for covering mushroom-beds. Beds made in this way in the autumn and winter months, and covered with a thick layer of litter and mats, seldom require any watering. The culture is not usually attempted in summer; the heat acting upon the littery covering giving rise to insects which destroy the mushrooms; but with care their culture is quite practicable at that season; in proof of which I may say that during the last week of July, 1868, I saw them gathered freely in a market-garden just beside the Gloucester Road Station of the Metropolitan Railway, where by using a coating of litter about a foot thick, and over that a layer of mats, it was possible to procure them in good condition throughout the hottest summer within memory. There are many acres of ground covered with beds made thus in the market-gardens round London.

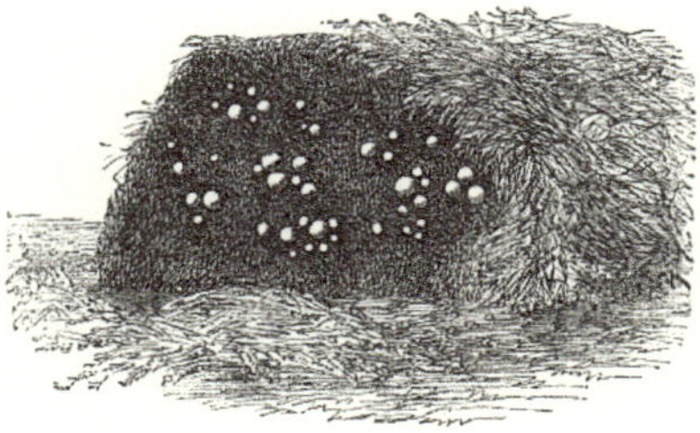

Figure 27 Uncovered end of mushroom-bed in Paris market-garden. January, 1867.

We will next turn to the culture of the mushroom in the open air near Paris. In old times the market-gardeners there used to grow it amongst their ordinary crops with great profit, but since the champignonnistes cultivate it under no danger from cold in the caves, the market-gardeners, who used to raise it to a great extent in the open air, do so now in a less degree. They begin with the preparation of the manure, and collect that of the horse for a month or six weeks before they make the beds; this they prepare in some firm spot of the market-garden, and take from it all rubbish, particles of wood, and miscellaneous matters; for, say they, the spawn is not fond of these bodies. After sorting it thus, they place it in beds two feet thick, or a little more, pressing it with the fork. When this is done the mass or bed is well stamped, then thoroughly watered, and finally again pressed down by stamping. It is left in this state for eight or ten days, by which time it has begun to ferment, after which the bed ought to be well turned over and re-made on the same place, care being taken to place the manure that was near the sides of the first-made bed towards the centre in the turning and re-making. The mass is now left for another ten days or so, at the end of which time the manure is about in proper condition for making the beds that are to bear the mushrooms. Little ridge-shaped beds—about twenty six inches wide and the same in height—are then formed in parallel lines at a distance of twenty inches one from the other.

In a market-garden they may stretch over a considerable extent, their length being determined by the wants of the grower. The beds once made of a firm, close-fitting texture, the manure soon begins to warm again, but does not become unwholesomely hot for the spread of the spawn. When the beds have been made some days, the cultivator spawns them, having of course ascertained beforehand that the heat is genial and suitable. Generally the spawn is inserted within a few inches of the base, and at about thirteen inches apart in the line. Some cultivators insert two lines, the second about seven inches above the first. In doing so, it would of course be well to make the holes for the spawn in an alternate manner. The spawn is inserted in flakes about the size of three fingers, and then the manure is closed in over, and pressed firmly around it. This done, the beds are covered with about six inches of clean litter. Ten or twelve days afterwards the growers visit the beds, to see if the spawn has taken well. When they see the white filaments spreading in the bed they know that the spawn has taken; if not, they take away the spawn they suppose to be bad and replace it with better. But, using good spawn, and being practised hands at the work, they rarely fail in this particular; and when the spawn is seen spreading well through the bed, then, and not before, they cover the beds with fresh sweet soil to the depth of about an inch or so. For cover, the little pathway between the beds is simply loosened up, and the rich soil of the market-garden applied equably, firmly, and smoothly with a shovel. With these open-air beds they succeed in getting mushrooms in winter. A covering of abundance of litter is put on immediately after the beds are earthed, and kept there as a protection. They have not long to wait till the beds are in full bearing, and when they are in that state it is thought better to examine and gather from them every second day, or even every day where there are many beds. And thus they grow excellent mushrooms, and in great quantity, all the further attention required being to renew the covering when it gets rotten, and an occasional watering in a very dry season.

Of course this kind of cultivation is perfectly practicable in private gardens—where, however, I have not yet seen it carried out. Where there is a mushroom-house or empty shed in which mushrooms may be grown, there would be less occasion to pursue it, but there are many places in which no such conveniences exist. In any case it is desirable that gardeners generally should know to what a large extent this phase of the culture is pursued round London and Paris, and how simply it is done. Instead of mats, it would be an improvement to cover the beds with tarpaulin or some other cheap material that would keep out the wet.

CHAPTER VIII

CULTURE IN GARDENS, ETC., WITH OTHER CROPS IN THE OPEN AIR

THIS is a phase of culture which may be pursued to great advantage in every private garden, almost without cost and attention. The low ridge-like hotbeds, for example, made for both long and short prickly cucumbers, gourds, marrows, &c., are admirably suited for growing a crop of mushrooms under the leaves of the subjects for which they were made. If the spawn be inserted soon after the beds are made, or at any convenient time in early summer, the beds will come into bearing in due course. Perhaps they may do so when mushrooms are found abundantly in the fields; but there are thousands of persons possessing gardens who have no fields in which to cull mushrooms, and who would like to gather them fresh in summer or autumn, if they could not afford to grow them in any covered structure in winter. And this is but one way in which they may be grown with summer garden crops, as will appear from the following communication, by Mr. Ayres, to the *Field*:—

"The finest crop and the best mushrooms I ever saw were grown in the open ground, and without any protection at all. I will tell you how it happened. Some years back I had the charge of the garden of a noted hunting establishment in Northamptonshire, one of the aids to success being that the manure of an average of nearly fifty highly-fed horses went to the garden, the owner remarking that, whatever other things I might run short of, there would be plenty of 'muck.' Well, the best of the hunters during the summer were soiled in loose boxes, principally under cover, and in these boxes the manure was allowed to accumulate until it began to grow too hot for the feet of the horses; then it was indispensable that it should be removed. About midsummer it so happened that nearly three acres of ground had been cleared of the spring crop, spinach, early peas, beans, &c., and I had determined to devote the whole plot to winter brassicas, broccoli, Brussels sprouts, &c. The ground was brashy and very poor, and consequently I determined to clear the boxes and put the whole of the manure upon it. It was carted away so rich in ammonia that the men who loaded it shed tears, not from sentiment, but from compulsion; and when the manure was spread upon the surface it was nothing less than a foot thick—so thick, that the proprietor said it was impossible for it to be dug into the ground. However, clearing a trench at one end of the piece, thirty inches wide and nearly a foot deep, the subsoil was broken up with strong steel forks, and upon that the dung covering the next strip was placed, and covered with the surface soil of the next trench; and so the work proceeded until the manure was put out of sight. I may remark that the dung, especially that around the walls, contained evidence of being strongly impregnated with mushroom spawn, though this was not regarded as being likely to produce a crop of the esculent. A soaking rain falling, the ground was immediately planted with brassicas, which grew as if they could not help growing—and in fact they could not.

"We had not planted for mushrooms, nor were mushrooms expected; but, walking round one morning early in September, a bunch of splendid fellows presented themselves, so large and thick and solid, that when I took them in for breakfast my *chef de cuisine* and 'better half' had grave doubts as to whether they were 'the real thing.' However, they were eaten, and the present writing is a proof that they did not poison me. Returning to the plot, I found the bunch gathered was not a solitary one—on the contrary, the ground was literally paved with mushrooms, many of them so large that bushels were gathered for ketchup within a few hours; while the retainers of a large establishment, down to the lowest labourer, were in a fortnight positively sick of them, and cartloads rotted upon the ground.

"The evidence of this unexpected success demonstrated two things—first, that if the ground is freely manured with *fresh*dung from well-fed horses, mushrooms are almost sure to be produced; and, secondly, that the more the ground is covered with the foliage of plants, the more certain will be the crop. Thus we found more mushrooms under savoys and broccoli than under Brussels sprouts—the former no doubt protecting the crop from heavy drenches of rain, which we know are very injurious to the mushroom crop. Since this example of mushroom-growing turned up, nearly fifteen years ago, I have frequently concentrated the fresh manure under a row of savoys or broccoli, throwing in at the same time a dust of mushroom spawn or the dung of a spent mushroom bed; and, except in very wet seasons, I have rarely failed to have a fine supply during the months of September and October. One point of success I believe to be essentially necessary, and that is, that water shall have a free passage through the ground at all times; hence the necessity of trenching the ground, if you expect mushrooms as well as brassicas."

Even in gardens where mushrooms are well grown in enclosed structures such results in early autumn will often be desirable; while in numbers of places where there are few or no opportunities of gathering them in abundance under other circumstances, crops in the garden will be very welcome. Therefore utilise the old mushroom-beds!

CHAPTER IX

MUSHROOM CULTURE IN PASTURES, ETC

NOTWITHSTANDING the extreme abundance of the common mushroom in the meadows and pastures of the British islands, and probably in similar positions all the world over, it is scarce in many situations, and, it may be, not a few persons would be willing to make it of more frequent occurrence in their fields. There is an opinion not uncommon that this cannot be done; that the mushroom is, to a great extent, a creature of chance, and that it cannot be cultivated. This is not a philosophical notion: there can be no doubt that the mushroom has to abide the results of the struggle for life as well as any other species of plant. Considering that we have taken the spawn from the fields and cultivated it with great success in all sorts of positions, none of which it could ever inhabit naturally, it is absurd to suppose that we cannot induce it to grow in positions exactly similar to its native habitat. Found in open, sunny meadows and pastures, and avoiding the shade of trees, it is grown, as we have seen, in dark and deep mines; yet people suppose it cannot be grown in those pastures in which it happens not to be found. It is erroneously inferred that there is something in its constitution or habit which causes it to occur in certain spots exclusively; but as well might we say this of any other plant. We know well that hundreds of native plants are hardy enough to grow almost anywhere, yet how many of them are but locally distributed and rare! Again, many plants are weeds in one district and unknown in another, perhaps, neighbouring one.

As the Rev. M. J. Berkeley remarks:—"It is almost useless to advert to the notion, though a very common one, which would regard these productions as the creatures of chance or of a happy concurrence of circumstances favourable to their growth from inorganic elements. It is true they often occur in unexpected situations, and from their extreme rapidity of development seem as if they could not have originated from anything like seed. But, as accurate inquiry has now thrown much light on the mystery in which the origin of intestinal worms was lately involved, so the phenomena which attend the growth of fungi are gradually receiving light, and they are found to follow essentially the same laws as more perfect vegetables." It is, in fact, quite fair to conclude that mushrooms, like most other plants, occupy but a small space in the vast expanse of soil and site which are naturally adapted for their growth. I read in a gardening journal that "it is impossible to command a crop of out-door mushrooms." I am positive that it can be done with almost as much certainty as any other crop, provided we take into consideration certain conditions. Of course, we must remember its natural wants; the more we do so, the more certain of success we may be. We know that it grows most abundantly in rich, upland pastures where water does not lie, associated with the meadow foxtail, meadow and hard fescue and cock's-foot grasses, clovers, cowslips, daisies, yarrow, &c., and also with the thistles (*Cnicus lanceolatus* and *C. arvensis*), and other plants fond of similar soils. We know that it is rarely found where the marsh plume-thistle (*Cnicus palustris*), tufted hair-grass, and other marsh grasses and plants abound, and from the presence or absence of these plants we may easily make up our minds as to the positions that suit it best. Now, it has long since been proved in gardens that it is quite possible to cultivate plants to a much higher degree of perfection than they ever attain in a wild state, under conditions entirely different, and it is not improbable that we should be able to grow the common mushroom in soils and positions far removed from those in which it naturally occurs. But there is no occasion for anything of the kind. It loves well-drained and dry pastures and meadows, and is not the country covered with such?

After selecting the position in which we wish to propagate mushrooms, and no moderately dry pastureland need be without them, the next thing to consider is the providing of the spawn. Hitherto this has probably been the great difficulty. When nearly 20*l.* worth of mushroom spawn was annually used in the mushroom-houses of a large garden, the expense necessary to spawn a large pasture might well alarm the richest of mushroom-loving landholders; but there is not the slightest occasion for purchasing the spawn for this purpose. Every farmer and country gentleman can make it as easily as, or more easily than, the spawn-manufacturer, without any expense or inconvenience, the essential thing being a quantity of rather short stable-manure.

Where this is gathered in large heaps it will be easy to obtain the requisite materials at once. Where it is not so, a few loads of stable manure unmixed with long straw may be thrown together in the open air and prepared for the purpose. There is no occasion to place it in a shed of any kind, though if there be one at hand so much the better. If prepared in the open air it should be on a dry place; the materials should be subjected to exactly the same preparation as when used for making a mushroom-bed, before described. They should be made into a potato-pit-shaped bed, and spawned in the usual manner. For this spawning it is of course necessary to obtain a little spawn, whether home-made or bought from the seedsman, or found in what the French call "a virgin condition" in the dunghill. In any case it will not be found difficult to spawn one or more beds in this way, particularly as there is nothing to prevent people drying as much home-made spawn at one time as will suffice for a year or more. The spawn should be allowed to run through this bed, which should be covered with a slight sprinkling of earth, and beaten pretty firm. When it has penetrated through the bed, it should, just before it arrives at a bearing condition, be ready to be used as spawn. The number of beds to be spawned in this way may be limited according to the extent of ground on which it is proposed to grow the mushrooms. This spawn may be inserted in the meadows in early summer, the most suitable time is in genial weather in May, and the spawn should be inserted in holes from six to ten feet apart.

The most expeditious and best way of inserting it is that termed T-planting, striking the spade in the line represented by the perpendicular of the T, and then in the horizontal one on the top, pressing the spade back when in the last position, so as to readily admit of the insertion of one or more pieces of spawn. The kind of spawn made as I have recommended usually falls into small pieces, more likely to impregnate the earth quickly than the stiff, brick-like pieces of nursery spawn. The ground, after the insertion of the spawn, should be pressed firm with the foot. As to the depth at which the spawn should be deposited, it would be better not to put it at any given depth, but so that while one piece of a flake may be at a depth of six inches or nearly so, others may touch the very surface. This, it need hardly be pointed out, would allow of the spawn vegetating at the depth and temperature most congenial to it. It would be most desirable to spawn at slightly different times, and, if possible, with different samples of spawn: thus, for example, it would be well to use a mixture of old and dried spawn with that taken fresh from one of the beds alluded to. If this were not convenient, some part of the large bed of spawn might be laid by to dry, and used a week or two afterwards. Probably the most economical way of doing this on a large scale would be by employing a number of boys, guided by an experienced workman.

It is scarcely desirable to attempt the culture in kept lawns, as no matter how suitable these are for it, the appearance of a large crop of mushrooms would have anything but a tendency to beautify the carpet of turf, and would probably become offensive from their odour.

The preceding refers to the cultivation of mushrooms in pastures, meadows, &c. There is not the slightest reason why a similar course of culture would not succeed in fields amongst green crops. As large crops of mushrooms have been produced in gardens under broccoli, &c., there is no reason whatever why they might not be grown in the same manner under field-turnips, mangold-wurtzel, &c. The spawn which could be so easily prepared by any farmer, could be readily inserted in the sides of the drills in which these crops are usually grown, the slight elevation of which, by preserving the spawn from excessive wet, will favour its development, and it would take possession of, and impregnate the manure in the drill. In fact, prodigious quantities might be raised in this and similar ways, with but little trouble; and should the fields be afterwards laid down, as is not uncommonly the case, the pasture or meadow would probably become a regular mushroom-ground.

CHAPTER X

THE COMMON MUSHROOMS

Agaricus campestris (True Meadow Mushroom)

THE common meadow mushroom varies considerably, but, "common to all are a fleshy *pileus*, which is sometimes smooth, sometimes scaly, in colour white, or of different shades of tawny, fuliginous, or brown; *gills* free, at first pallid, then flesh-coloured, then pink, next purple, at length tawny-black; the *stem* white, full, firm, varying in shape, furnished with a white persistent ring; the *spores* brown-black, and a volva which is very *fugacious*."—*Badham's Esculent Funguses of England.*

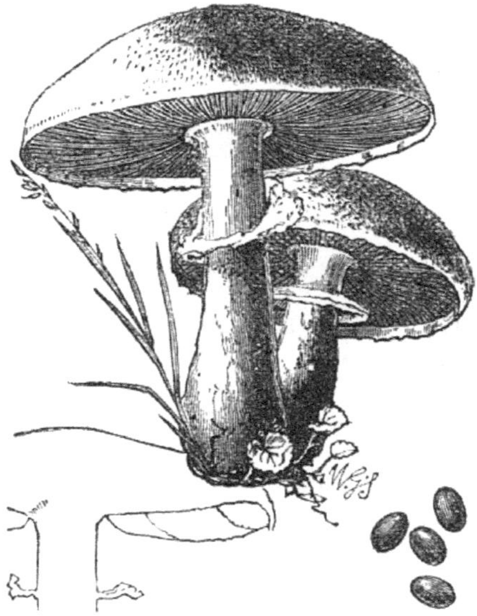

Figure 28 *Agaricus campestris* (the True Meadow Mushroom). Pastures, autumn; colour, white or pale brown; gills, salmon, at length black; diameter, 3 to 6 inches. The spores are magnified 700 diameters.

There is scarcely any one in England who does not feel himself competent to decide on the genuineness of a mushroom; its pink gills easily distinguish it from a kindred fungus, *Ag. arvensis*, the gills of which are of a flesh-coloured grey, and out of the pickings of ten thousand hands, a mistake is of rare occurrence; and yet no fungus presents itself under such a variety of forms, or such singular diversities of aspect! The inference is plain; less discrimination than that employed to distinguish this would enable anyone who should take the trouble to recognise at a glance many of those esculent species, which every spring and autumn fill our plantations and pastures with plenteousness. Neither is this left to be a mere matter of inference; it is corroborated in a singular manner by what takes place at Rome; there, whilst many hundred baskets of what we call toadstools are carried home for the table, almost the only one condemned to be thrown into the Tiber, by the inspector of the fungus market, is our own mushroom; indeed, in such dread is this held in the Papal States, that no one knowingly would touch it. "It is reckoned one of the fiercest imprecations," writes Professor Sanguinetti, "amongst our lower orders, infamous for the horrible nature of their oaths, to pray that one may die of a *Pratiolo*;" and although it has been some years registered among the esculent funguses of Milan and Pavia (on the authority of Vittadini), it has not yet found its way into those markets. Mr. Worthington G. Smith, in his "Mushrooms and Toad-stools," qualifies this statement of Dr. Badham.

Agaricus campestris is not generally appreciated in Italy, and indeed is seldom eaten, and never appears in the markets, for the simple reason that there would be no sale for it. There is an edict in existence ordering certain fungi to be thrown into the Tiber, but it is now, and has long been altogether effete; and whilst there is an abundance of *A. Cæsareus* (avowedly the most delicious of all fungi) for the markets of Italy, it is not to be expected the consumption will be given up for another and little known species.

The Modes of Cooking this Species.—"The mushroom, having the same proximate principles as meat, requires, like meat, to be cooked before these become changed. The *Ag. campestris* may be prepared in a great variety of ways: they give a fine flavour to soup, and greatly improve beef-tea; where arrowroot and weak broths are distasteful to the patient, the simple seasoning of a little ketchup will frequently form an agreeable change. Some roast them, basting with melted butter and white (French) wine sauce. In patties and *vols-au-vent* they are equally excellent; in fricassees, as everybody knows, they are the important element of the dish. Roques recommends in all cases the removal of the gills before dressing, which though it secures a more elegant-looking *entremets*, is only flattering the eye at the expense of the palate."—*Badham.*

Agaricus arvensis (Horse-Mushroom)

"*Pileus* fleshy, obtusely conico-campanulate, then expanded, at first floccose, then smooth, even, or rivulose; *stem* hollow, with a floccose pith; *ring* broad, pendulous, double, the outer split in rays; *gills* free, wider in front, at first dirty white, then brown, tinged with pink."—*Berkeley's Outlines of British Fungology.*

Figure 29 *Agaricus arvensis* (Horse-Mushroom). Pastures, in autumn; colour, yellowish; gills pallid, at length black; diameter, 6 to 24 in.

"This species is very nearly allied to the meadow mushroom, and frequently grows with it, but it is coarser, and has not the delicious flavour. It is usually much larger, often attaining enormous dimensions; it turns a brownish yellow as soon as broken or bruised. The top in good specimens is smooth, and snowy white; the gills are not the pure pink of the meadow mushroom, but dirty brownish white, ultimately becoming brown-black. It has a big, ragged, floccose ring, and the pithy stem is inclined to be hollow. It is *the* species exposed for sale in Covent Garden Market. Indeed, after knowing the market for many years, I have rarely seen any other species there; when the true mushroom, however, *is* there, it is frequently mingled with horse mushrooms, which seems to show that the dealers do not know one from the other. In the wet days of autumn, children, idlers, and beggars go a few miles from town into the meadows to gather whatever they can find in the mushroom line; they then bring their dirty stock to market, where it is sold to fashionable purchasers; stale, vapid, and without taste—unless it be a bad one.

"When young and fresh, the horse mushroom is a most desirable addition to the bill of fare: it yields an abundant gravy, and the flesh is firm and delicious. It is a valuable plant when freshly gathered, but when stale it becomes tough and leathery, and without aroma or juice.

"There is a curious, large, brown, hairy variety, of rather uncommon occurrence, similar to the hairy variety of the meadow mushroom, the *A. villaticus* of Dr. Badham. It is a splendid form, but, I think, very rare. I have only seen it once.

"Many country-folk readily distinguish the meadow from the horse mushroom, and show antipathy to the latter, although they are always willing to put it into the jar as one of the ingredients of ketchup. Opinions appear to differ greatly regarding the excellence of this species. Mr. Penrose writes:—'I think young, and especially button, specimens of this very indigestible; until they are well opened out, they are unfit for use.' Such, however, I must say, is not my experience of button specimens.

"There is a strong odour attached both to the fungus and the spawn, the ground just below the surface being frequently white with the latter; or if horse-dung be kicked aside in a rich meadow frequented by graminivorous animals, the earth will frequently present a snowy whiteness from the spawn of this species, from which the young individuals may be seen springing up.

"I once saw a sheep eat a large specimen with great apparent gusto, although the fungus was full of maggots."—*Worthington G. Smith.*

CHAPTER XI

MODES OF COOKING THE COMMON MUSHROOMS

THE following modes of cooking mushrooms may prove useful to some:—

To Stew Mushrooms.—Trim and rub clean half a pint of large button mushrooms; put into a stew-pan two ounces of butter, shake it over the fire till thoroughly melted; put in the mushrooms, a tea-spoonful of salt, half as much pepper, and a blade of mace pounded; stew till the mushrooms are tender, then serve them on a hot dish. They are usually sent in as a breakfast dish, thus prepared in butter.

Mushrooms à la Crême.—Trim and rub half a pint of button mushrooms, dissolve two ounces of butter rolled in flour in a stew-pan; then put in the mushrooms, a bunch of parsley, a tea-spoonful of salt, half a tea-spoonful each of white pepper and of powdered sugar, shake the pan round for ten minutes, then beat up the yolks of two eggs, with two table-spoonfuls of cream, and add by degrees to the mushrooms; in two or three minutes you can serve them in the sauce.

Mushrooms on Toast.—Put a pint of mushrooms into a stew-pan, with two ounces of butter rolled in flour; add a tea-spoonful of salt, half a tea-spoonful of white pepper, a blade of mace powdered, and half a tea-spoonful of grated lemon; stew till the butter is all absorbed, then add as much white *roux* as will moisten the mushrooms; fry a slice of bread in butter, to fit the dish, and as soon as the mushrooms are tender serve them on the toast.

To Pot Mushrooms.—The small open mushrooms suit best for potting. Trim and rub them; put into a stew-pan a quart of mushrooms, three ounces of butter, two tea-spoonfuls of salt, and half a tea-spoonful of Cayenne and mace mixed, and stew for ten or fifteen minutes, or till the mushrooms are tender; take them carefully out and drain them perfectly on a sloping dish, and when cold press them into small pots, and pour clarified butter over them, in which state they will keep for a week or two. If required to be longer preserved, put writing paper over the butter, and over that melted suet, which will effectually preserve them for many weeks, if kept in a dry, cool place.

To Pickle Mushrooms.—Select a number of small, sound, pasture mushrooms as nearly as possible alike in size; throw them for a few minutes into cold water; then drain them; cut off the stalks, and gently rub off the outer skin with a moist flannel dipped in salt; then boil the vinegar, adding to each quart two ounces of salt, half a nutmeg sliced, a drachm of mace, and an ounce of white pepper-corns; put the mushrooms into the vinegar for ten minutes over the fire; then pour the whole into small jars, taking care that the spices are equally divided; let them stand a day, then cover them.

Another Method.—In pickling mushrooms take the buttons only and while they are quite close, cut the stem off even with the gills and rub them quite clean. Lay them in salt and water for forty-eight hours, and then add pepper, and vinegar in which black pepper and a little mace have been boiled. The vinegar must be applied cold. So pickled they will keep for years.

Mushrooms en Ragoût.—Put into a stew-pan a little stock, a small quantity of vinegar, parsley, and green onions chopped up, salt, and spices. When this is about to boil, the mushrooms being cleaned, put them in. When done remove them from the fire, and thicken with yolks of eggs.

Mushrooms and Toast.—Peel the mushrooms, and take out the stems. Fry them over a quick fire. When the butter is melted take off the pan. Squeeze the juice of a lemon into it. Let the mushrooms fry again for some minutes. Add salt, pepper, spices, and a spoonful of water, in which a clove of garlic, having been cut into pieces, has soaked for half an hour; let it stew. When the mushrooms are done, make a thickening of yolks of eggs. Pour the mushrooms on bread fried in butter, and laid in the dish ready for them.

Mushrooms en Caisse.—Peel the mushrooms lightly, and cut them into pieces. Put them into cases of buttered paper, with a bit of butter, parsley, green onions, and shalots chopped up, salt and pepper. Dress them on the gridiron over a gentle fire, and serve in the cases.

Mushrooms à la Provençale.—Take mushrooms of good size. Remove the stems, and soak them in olive oil. Cut up the stems with a clove of garlic and some parsley. Add meat of sausages, and two yolks of eggs to unite them. Dish the mushrooms, and garnish them with the forcemeat. Sprinkle them with fine oil, and dress them in an oven, or in a *four de campagne*.

Baked Mushrooms.—Peel the tops of twenty mushrooms; cut off a portion of the stalks, and wipe them carefully with a piece of flannel dipped in salt. Lay the mushrooms in a tin dish, put a small piece of butter on the top of each, and season them with pepper and salt. Set the dish in the oven, and bake them from twenty minutes to half an hour. When done, arrange them high in the centre of a very hot dish, pour the sauce round them, and serve quickly, and as hot as you possibly can.

Mushrooms au Gratin.—Take twelve large mushrooms about two inches in diameter, pare the stalks, wash, and drain the mushrooms on a cloth; cut off and chop the stalks. Put in a quart stew-pan an ounce of butter and half an ounce of flour; stir over the fire for two minutes; then add one pint of broth; stir till reduced to half the quantity. Drain the chopped stalks of the mushrooms thoroughly in a cloth; put them in the sauce with three table-spoonfuls of chopped and washed parsley, one table-spoonful of chopped and washed shalot, two pinches of salt, a small pinch of pepper; reduce on a brisk fire for eight minutes, put two table-spoonfuls of oil in a *sauté* pan; set the mushrooms in, the hollow part upwards; fill them with the fine herbs, and sprinkle over them lightly a table-spoonful of raspings; put in a brisk oven for ten minutes, and serve.

Mushroom Soup.—Take a good quantity of mushrooms, cut off the earthy end, and pick and wash them. Stew them with some butter, pepper, and salt in a little good stock till tender; take them out, and chop them up quite small; prepare a good stock as for any other soup, and add it to the mushrooms and the liquor they have been stewed in. Boil all together, and serve. If white soup be desired, use the white button mushrooms, and a good veal stock, adding a spoonful of cream or a little milk, as the colour may require.

The following "family receipts" have been communicated by a friend:

Clean a dozen or so of medium-size, place two or three ounces of nice clean beef-dripping in the frying-pan, and with it a table-spoonful or more of nice beef gravy. Set the pan on a gentle fire, and as the dripping melts place in the mushrooms, adding salt and pepper to taste. In a few minutes they will be cooked, and being soaked in the gravy and served upon a hot plate, will form a capital dish. In the absence of gravy, a *soupçon* of "extractum carnis" may be substituted.

Mushrooms with Bacon.—Take some full-grown mushrooms, and having cleaned them, procure a few rashers of nice streaky bacon, and fry it in the usual manner. When nearly done, add a dozen or so of mushrooms, and fry them slowly until they are cooked. In this process they will absorb all the fat of the bacon, and with the addition of a little salt and pepper, will form a most appetising breakfast relish.

Mushroom Stems, if young and fresh, make a capital dish for those who are not privileged to eat the mushrooms. Rub them quite clean, and after washing them in salt and water, slice them to the thickness of a shilling, then place them in a saucepan with sufficient milk to stew them tender; throw in a piece of butter and some flour for thickening, and salt and pepper to taste. Serve upon a toast of bread, in a hot dish, and add sippets of toasted bread. This makes a light and very delicate supper dish, and is not bad sauce to a boiled fowl.

CHAPTER XII

SOME OF THE MOST COMMON AND USEFUL EDIBLE FUNGI

"Whole hundredweights of rich, wholesome diet rotting under the trees; woods teeming with food, and not one hand to gather it; and this, perhaps, in the midst of potato-blight, poverty, and all manner of privations, and public prayers against imminent famine."

Dr. Badham.

VALUABLE as is the common mushroom, it is indisputable that not a few other kinds are also capable of affording excellent food. Therefore, figures are given of the most prevalent, useful, and easily recognised kinds of edible fungi, as well as of the common mushrooms of our gardens and markets. These figures have been admirably drawn by Mr. W. G. Smith, and are accompanied by what seemed the most satisfactory accounts of the characters and properties that are obtainable. The spores which accompany the figures are uniformly enlarged seven hundred diameters.

Marasmius oreades (Fairy-ring Champignon)

Pileus smooth, fleshy, convex, subumbonate, generally more or less compressed, tough, coriaceous, elastic, wrinkled; when water-soaked, brown; when dry, of a buff or cream-colour, the umbo often remaining red-brown, as if scorched; *gills*free, distant, ventricose, of the same tint as the pileus, but more pale; *stem* equal, solid, twisted, very tough and fibrous, of a pale silky-white colour.

Figure 30 1.*Marasmius oreades* (Fairy-ring Champignon). Pastures, roadsides, and downs, in the autumn; colour, pale buff; *gills broad and far apart*; diameter, 1 to 2 inches.

Figure 30 2. *Marasmius urens* (False Champignon). Woods and pastures in the autumn; colour, pale buff; *gills narrow and crowded together*; diameter, ½ inch to 1½ inches.

The fairy-ring agaric is a valuable little fungus, and common on almost every lawn. In hilly pastures it generally appears in broad brown patches, either circular or forming a portion of a circle.

M. urens, the most acrid of all allied funguses, usually grows in woods, though sometimes in the fairy-ring. However, its flat top and narrow crowded gills cause it to be readily distinguished anywhere.

Opinions on the Merits of Marasmius oreades as an edible Fungus.—"On the Continent this species has long been considered edible, but on account of its coriaceous texture it is dried and employed in the form of powder, to season various made-dishes."—*Dr. Greville.*

"The common fairy-ring champignon is the best of all our funguses, yet there is scarcely one person in a thousand who dare venture to use it. With common observation no mistake need be made with regard to it. It has an extremely fine flavour, and makes perhaps the very best ketchup that there is."—*Rev. M. J. Berkeley.*

"An excellent flavour, as good as that of most funguses."—*Dr. Badham.*

Modes of Cooking Marasmius oreades.—General Use.—"Cut in small pieces and seasoned it makes an excellent addition to stews, hashes, or fried meats, but it should only be added a few minutes before serving, as the aroma is dissipated by over cooking. It is the mushroom used in the French *à la mode* beef-shops in London."—*Dr. Badham.*

When stewed, the champignons require rather longer time to ensure their being made perfectly tender. They are readily dried by removing the stems from the fungus, threading them on a string, and hanging them up in a dry airy place. "When dried, it may be kept for years without losing any of its aroma or goodness, which, on the contrary, becomes improved by the process, so as, in fact, to impart more flavour to the dish than would have been imparted by the fresh fungus; though it is not to be denied that the flesh then becomes coriaceous (or tough), and less easy of digestion."—*Dr. Badham.*

Champignon Powder.—Put the champignons in a stew-pan with a little mace and a few cloves, and a sprinkling of white pepper. Simmer, and shake constantly to prevent burning, until any liquor that may exude is dried up again. Dry thoroughly in a warm oven until they will easily powder. Put the dried agaric, or the powder, into wide-mouthed glass bottles, and store in a dry place. It will keep any length of time. A tea-spoonful added to any soup, or gravy, or sauce, just before the last boil is given, will produce a very fine mushroom flavour.

Pickled Champignons.—Collect fresh buttons of the fairy-ring agaric and use them at once. Cut off the stems quite close and throw each one as you do so into a basin of water in which a spoonful of salt has been put. Drain them from it quickly afterwards, and place them on a soft cloth to dry. For each quart of buttons thus prepared, take nearly a quart of pale white wine vinegar, and add to it a heaped tea-spoonful of salt, half an ounce of whole white pepper, an ounce of ginger-root bruised, two large blades of mace, and a fourth of a salt-spoon of cayenne pepper tied in a small piece of muslin. When this pickle boils throw in the agarics and boil them in it over a clear fire moderately fast, from six to nine minutes. When tolerably tender put them into *warm* wide-mouthed bottles, and divide the spice equally amongst them. When perfectly cold, cork well, or tie skins and paper over them. Store in a dry place, and keep out the frost.

Full-sized champignons may be pickled exactly in the same way, but will require longer boiling, until indeed they become tender.—*Modified from Miss Acton.*

Champignons quickly Pickled.—Place the prepared buttons in bottles with a blade of mace, a tea-spoonful of pepper-corns, and a tea-spoonful of mustard seed in each, and cover with the strongest white wine pickling vinegar boiling hot. Cork or tie down as before, but do not expect them to keep above three months.

Agaricus procerus (the Parasol Agaric)

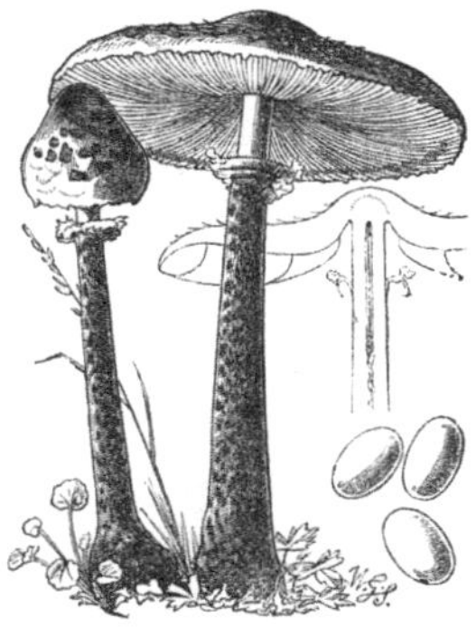

Figure 31 *Agaricus procerus* (Scaly Mushroom). Pastures, &c., in autumn; colour, pale brownish buff; diameter, 5 to 12 inches.

Pileus fleshy, ovate when young, then campanulate, and afterwards expanded and umbonate (blunt pointed), from three to seven inches across. Cuticle more or less brown, entire over the umbo, but torn into patches, or scales which become more and more separated as they approach the margin. Flesh white. *Gills* unconnected with the stem, fixed to a collar on the pileus surrounding its top. *Ring* persistent, loose on the stem. *Stem* six or eight inches high, tapering upwards from a pear-like bulb at the root, hollow with a loose pith, whitish brown, but more or less variegated with small and close-pressed scales.

Whenever an agaric on *a long stalk*, enlarged *at the base*, presents *a dry cuticle* more or less *scaly*, a darker coloured *umbonated top*, *a moveable ring*, and *white* gills, it must be *Agaricus procerus*, the parasol agaric, and it may be gathered and eaten without fear. When the whitish flesh of this agaric is bruised it shows a light reddish colour.

There are but two other agarics that at all resemble it, and both are edible. One about the same size is *Agaricus rachodes*. It is not generally considered so good in flavour as *A. procerus*. Mrs. Hussey, however, says plainly, "If *Agaricus procerus* is the king of edible funguses, *Agaricus rachodes* is an excellent viceroy." The other is the *Agaricus excoriatus*, a very much smaller fungus, with a more slender habit, a shorter stem, and no true bulb at the base. This elegant little fungus is also very good eating.

The parasol agaric has a very wide range of growth. It is a common fungus, and is in *high request all over the Continent*.

Opinions on the Merits of Agaricus procerus as an Edible Fungus.—"A most excellent mushroom, of a delicate flavour, and it must be considered a most useful species."—*The Rev. M. J. Berkeley.*

"Were its excellent qualities better known here, they could not fail to secure it a general reception into our best kitchens, and a frequent place among our side dishes at table."—*Dr. Badham.*

"If once tried, it must please the most fastidious."—*Worthington G. Smith.*

There can be no question but that, when young and quickly grown, the parasol agaric is a delicious fungus. It has a light and delicate flavour without the heavy richness which belongs to the ordinary field mushroom. The writer has prevailed on many persons to try it; all without exception have liked it, many have thought it quite equal, and some have proclaimed it superior, to the common mushroom.

Modes of Cooking the Agaricus procerus.—Broiled Procerus.—Remove the scales and stalks from the agarics, and broil lightly over a clear fire on both sides for a few minutes; arrange them on a dish over fresh-made, well-divided toast; sprinkle with pepper and salt, and put a small piece of butter on each; set before a brisk fire to melt the butter, and serve up quickly.

If the cottager would toast his bacon over the broiled mushrooms, the butter would be saved.

Agarics delicately Stewed.—Remove the stalks and scales from young half-grown agarics, and throw each one as you do so into a basin of fresh water slightly acidulated with the juice of a lemon, or a little good vinegar. When all are prepared, remove them from the water, and put them into a stew-pan with a very small piece of fresh butter. Sprinkle with white pepper and salt, and add a little lemon-juice; cover up closely, and stew for half an hour. Then add a spoonful of flour, with sufficient cream, or cream and milk, until the whole has the thickness of cream. Season to taste, and stew again gently until the agarics are perfectly tender. Remove all the butter from the surface, and serve in a hot dish, garnished with slices of lemon.

A little mace, nutmeg, or ketchup may be added; but there are those who think that spice spoils the mushroom flavour.

Cottager's Procerus Pie.—Cut fresh agarics in small pieces, and cover the bottom of a pie-dish. Pepper, salt, and place them on small shreds of fresh bacon, then put in a layer of mashed potatoes, and so fill the dish, layer by layer, with a cover of mashed potatoes for the crust. Bake well for half an hour, and brown before a quick fire.

A la Provençale.—"Steep for two hours in some salt, pepper, and a little garlic; then toss in a small stew-pan over a brisk fire, with parsley chopped, and a little lemon-juice."—*Dr. Badham.*

Agaric Ketchup.—Place agarics of as large a size as you can procure, but which are not worm-eaten, layer by layer, in a deep pan, sprinkling each layer as it is put in with a little salt. The next day stir them well up several times, so as to mash and extract their juice. On the third day strain off the liquor, measure, and boil for ten minutes, and then to every pint of the liquor add half an ounce of black pepper, a quarter of an ounce of bruised ginger-root, a blade of mace, a clove or two, and a tea-spoonful of mustard-seed. Boil again for half an hour; put in two or three bay leaves, and set aside till quite cold. Pass through a strainer, and bottle; cork well, and dip the ends in resin. A very little Chili vinegar is an improvement, and some add a glass of port wine, or a glass of strong ale to every bottle.

Care should be taken that the spice is not added so abundantly as to overpower the true flavour of the agaric. A careful cook will keep back a little of the simple boiled liquor to guard against this danger: a good one will always avoid it. "Doctors weigh their things," said a capital cook, "but I go by taste." But then, like poets, good cooks of this order must be born so; they are not to be made.

Coprinus comatus (the Maned Agaric)

Pileus cylindrical, obtuse, campanulate, fleshy in the centre, but very thin towards the margin. The external surface soon torn up into fleecy scales, with the exception of a cap at the top. *Gills* free, linear, and crowded. Quite white when young, becoming rose-coloured, sepia, and then black, from the margin upwards. They then expand quickly, curl up in shreds, and deliquesce into a black inky fluid which stains the ground. *Stem* of a pure white, four to five inches high, contracting at the top, and bulbous at the base; hollow, fibrillose, stuffed with a light cottony web. The bulb is solid and rooting, the ring is movable.

Figure 32 *Coprinus comatus* (Maned Mushroom). Pastures, parks, and roadsides, summer and autumn; colour, snow-white; height, 5 to 12 inches.

This very elegant agaric has also been called *Ag. cylindricus*, Schœff; *Ag. typhoides*, Bull; and *Ag. fimetarius*, Bolt. It is common throughout the summer and autumn months, on road-sides, pastures, and waste places. It is extremely variable in size. Its general appearance is so distinct and striking, that it cannot possibly be mistaken for any other agaric. It grows so abundantly on waste ground in the dwellings and farm-yards that it may be, says Dr. Bull, called the "agaric of civilization;" and for both these reasons it is most valuable as an edible agaric. If its merits were known, it would be eaten as freely as the common field mushroom.

"The maned mushrooms," Miss Plues has well said, "grow in dense clusters, each young plant like an attenuated egg, white and smooth. Presently some exceed the others in rapidity of growth, and their heads get above the ground, the stem elongates rapidly, the ring falls loosely round the stem, the margin of the pileus enlarges, and the oval head assumes a bell-shape; then a faint tint of brown spreads universally or in blotches over the upper part of the pileus, and the whiteness of its gills changes to a dull pink. A few more hours and the even head of the pileus has split in a dozen places, the sections curl back, melt out of all form into an inky fluid, and on the morrow's dawn a black stain on the ground will be all that remains. And so on with the others in succession."

Opinions on the Merits of Coprinus comatus as an Edible Fungus.—"Esculent when young."—*Berkeley.*

"Young specimens should be selected."—*Badham.*

"No despicable dish, though perhaps not quite equal to the common mushroom."—*M. C. Cooke.*

"If I had my choice, I think there is no species I should prefer before this one: it is singularly rich, tender, and delicious."—*Worthington G. Smith.*

Dr. M'Cullough, Dr. Chapman, Elmes Y. Steele, Esq., and some other members of the Woolhope Club, hold Mr. W. G. Smith's opinion as the result of considerable experience. It must be noted, however, that when too young this agaric is rather deficient in flavour, and its fibres tenacious. Its flavour is most rich, and its texture most delicate when the gills show the pink colour with sepia margins.

Modes of Cooking the Coprinus comatus.—The best and simplest method is to broil it and serve on toast in the ordinary way. It may be added also with great advantage to steaks and made-dishes, to give flavour and gravy.

Comatus Soup.—Take two quarts of white stock, and put in a large plateful of the maned agaric roughly broken out; stew until tender; pulp through a fine sieve; add pepper and salt to taste; boil and serve up hot. Two or three table-spoonfuls of cream will be a great improvement.

The agarics for this soup should be young, in order to keep its colour light and good. The maned agaric is recommended on all sides for making ketchup, but here, also, it should be quickly used, and the ketchup quickly made.

Agaricus gambosus (the True St. George's Mushroom)

Pileus thick and fleshy, convex at first, often lobed, becoming undulated and irregular, expanding unequally; the margin more or less involute, and at first flocculose; from three to four inches across; of a light yellow colour in the centre, fading to almost opaque white at the edges; it is soft to the touch; more or less tuberculated, and often presenting cracks. *Gills*yellowish-white, watery, narrow, marginate, annexed to the stem with a little tooth: they are very numerous and irregular, with many smaller ones interposed, "lying over each other like the plaits of a frill" (from 5 to 11, Vittadini). *Stem* firm, solid and white, swelling at the base in young specimens; but in older ones, though usually bulging, they are frequently of even size, and when in long grass they occasionally even taper downwards. This agaric is usually nearly white, smooth, soft, and firm, like kid leather to the touch, and, as Berkeley has happily said, "in appearance it very closely resembles a cracknel biscuit."

They grow in rings; have a strong smell, and appear about St. George's Day (April 23), after the rains which usually fall about the third week in April. They continue to appear for three or four weeks, according to the peculiarities of the season. They are usually to be found on hilly pastures in woodland districts.

The St. George's mushroom cannot well be mistaken for any other. The fact of its appearance at this early season, and growing so freely in rings, when so very few other funguses are to be found, is almost enough to distinguish it. It has, however, very distinctive characters in itself in the thickness of its pileus; the narrowness of its gills, which are very closely crowded together; and the solid bulging stem.

Figure 33 *Agaricus gambosus* (St. George's Mushroom). Pastures, *in the spring*; colour, cream; diameter, 4 to 6 inches.

The St. George's mushroom is not an uncommon agaric in this country, and where it does appear it is usually plentiful—a single ring affording generally a good basket full. It should be gathered when young, or it will be found grub-eaten, for no fungus is more speedily and more voraciously attacked by insects than this one.

Opinions on the Merits of Agaricus gambosus as an Edible Fungus.—"This rare and most delicious agaric, the*mouceron* of Bulliard, and the *Agaricus prunulus* of other authors, abounds on the hills above the valley of Staffora, near Bobbio, where it is called *Spinaroli*, and is in great request; the country people eat it fresh in a variety of ways, or they dry and sell it at from twelve to sixteen francs a pound."—*Letter from Professor Balbi to Persoon.*

"The most savoury fungus with which I am acquainted ... and which is justly considered over almost the whole continent of Europe as the *ne plus ultra* of culinary friandise.

"The *prunulus* (*gambosus*) is much prized in the Roman market, where it easily fetches, when fresh, thirty baiocchi—*i.e.*, fifteen pence per pound—a large sum for any luxury in Rome. It is sent in little baskets as presents to patrons, fees to medical men, and bribes to Roman lawyers."—*Dr. Badham.*

The *Agaricus gambosus* "is one that a person cannot well make any mistake about. It sometimes attains a large size, is excellent in flavour, and particularly wholesome."—*Rev. M. J. Berkeley.*

Mode of Cooking Agaricus gambosus.—"The best mode of cooking *Agaricus gambosus* is either to mince or fricassee it with any sort of meat, or in a *vol-au-vent*, the flavour of which it greatly improves; or simply prepared with salt, pepper, and a small piece of bacon, lard, or butter, to prevent burning, it constitutes of itself an excellent dish."—*Dr. Badham.* "Served with white sauce, it is a capital appendage to roast veal."—*Edwin Lees.* It may be broiled, stewed, or baked.

Breakfast Agaric.—Place some fresh-made toast, nicely divided, on a dish, and put the agarics upon it; pepper, salt, and put a small piece of butter on each; then pour on each one a tea-spoonful of milk or cream, and add a single clove to the whole dish. Place a bell-glass, or inverted basin, over the whole; bake twenty minutes, and serve up without removing the glass until it comes to the table, so as to preserve the heat and the aroma, which, on lifting the cover, will be diffused through the room. It dries very readily when divided into pieces, and retains most of its excellence. A few pieces added to soups, gravies, or made-dishes, give a delicious flavour.

Agaricus rubescens (Brown Warty Agaric)

Figure 34 *Agaricus rubescens* (Red-fleshed Mushroom). Woods, summer and autumn; colour, sienna-brown; diameter, 4 to 10 inches.

Pileus convex, then expanded, cuticle brown, scattered over with warts varying in size. Margin striate. *Gills* white, reaching the stem, and forming very fine decurrent lines upon it. *Ring* entire, wide and marked with striæ. *Stem* often scaly, stuffed, becoming hollow; when old, bulbous. Volva obliterated. The whole plant has a tendency to turn a sienna-red, or rust colour. This is very distinctly shown some little time after it has been bruised.

It is very common all through the summer and autumn months; indeed, one of the most abundant mushrooms; "and it is one of those species that a person with the slightest powers of discrimination may distinguish accurately from others."—*Badham.*

Opinions on the Merits of Agaricus rubescens as an Edible Fungus.—"A very delicate fungus, which grows in sufficient abundance to render it of importance in a culinary point of view."—*Badham.*

"From long experience I can vouch for its being not only wholesome, but, as Dr. Badham says, 'a very delicate fungus.'"—*F. Currey*, Editor of Dr. Badham's "Esculent Funguses."

Modes of Cooking the Agaricus rubescens.—It may be toasted, boiled, or stewed in the ordinary way.

Fried Rubescens.—Place the full-grown agarics in water for ten minutes, then drain, and having removed the warty skin, fry with butter, pepper, and salt. The ketchup made from *Agaricus rubescens* is rich and good. "As it grows freely, and attains a considerable size, it is very suitable for that purpose, quantity being a great desideratum in ketchup-making."—*Plues.*

Agaricus nebularis (Clouded Mushroom)

Figure 35 Agaricus nebularis (Clouded Mushroom). Woody places, in autumn; colour, cream, with slate-coloured top; diameter, 4 to 10 inches.

"*Pileus* from two and a half to five inches across; at first depresso-convex; when expanded, nearly flat or broadly subumbonate; never depressed; margin at first involute and pruinose; occasionally somewhat waved and lobed, but generally regular in form; smooth, viscid when moist, so that dead leaves adhere to it; grey, brown at the centre, paler towards the circumference. *Flesh* thick, white, unchanging. *Gills* cream-colour, narrow, decurrent, close, their margins waved, unequal, generally simple. *Stem* from two to four inches long, from a quarter of an inch to an inch thick; incurved at the base; not rooting, but attaching by means of a floccose down round its lower portion and for one-third of its length, a large quantity of dead leaves, by which the plant is held erect; subequal, more or less marked with longitudinal pits, firm externally, within of a softer substance. The *odour* strong, like that of curd cheese."—*Badham.*

"Common in certain places, but very rare near London. This species comes up late in the autumn on dead leaves in moist places, principally on the borders of woods. The gastronomic excellences of this species are well known. When gathered, it has a wholesome and powerful odour; and when cooked, the firm and fragrant flesh has a particularly agreeable and palatable taste."—*W. G. Smith.*

"The *Agaricus nebularis* requires but little cooking; a few minutes' broiling (*à la* Maintenon is best), with butter, pepper, and salt, is sufficient. It may also be delicately fried with bread crumbs, or stewed in white sauce. The flesh of this mushroom is perhaps lighter of digestion than that of any other."—*Badham.*

Lactarius deliciosus (Orange-milk Mushroom)

Figure 36 *Lactarius deliciosus* (Orange-milk Mushroom). Under fir-trees, in autumn; colour, brown-orange; milk at first orange, then green; diameter, 3 to 10 inches.

Pileus smooth, fleshy, umbilicate, of a dull rufous orange, turning pallid from exposure to light and air, but zoned with concentric circles of a brighter hue; margin smooth, at first involute, and then becoming expanded; from three to five inches across. Flesh firm full of orange-red milk, which turns green on exposure to the air, as does any part of the plant when bruised. *Gills* decurrent, narrow, each dividing into two, three several times from the stem to the edge of the pileus; of a dull yellow by reflected light, but being translucent, the red milk shines brightly through them. *Stem* from one to three inches high, slightly bent and tapering downwards; solid, becoming more or less hollow with age; short hairs at the base; sometimes pitted (scrobiculate).

There is no possibility of mistaking this fungus. It is the only one which has *orange-red milk*, and which *turns green when bruised.* These properties distinguish it at once from *Lactarius torminosus* or *necator*, the only fungus which in any way resembles it.

This acrid fungus (*Lactarius torminosus*) is somewhat similar in shape and size, and is also zoned. But the involute edges of the pileus are bearded with close hairs. It is of a much paler colour, and with gills of a dirty white. The milk, also, is white, acrid, and unchangeable in colour.

The Orange-milk agaric chiefly affects the Scotch fir-tree, and is generally to be found beneath the drip of the branches around the tree. It is also found in hedgerows occasionally, but is most abundant in plantations of Scotch fir or larch.

Opinions on the Merits of Lactarius deliciosus as an Edible Fungus.—"This is one of the best agarics with which I am acquainted, fully deserving both its name and the estimation in which it is held abroad, it reminds me of tender lambs' kidneys."—*Dr. Badham.*

"Very luscious eating, full of rich gravy, with a little of the flavour of mussels."—*Sowerby.*

"Cook them well, and you will have something better than kidneys, which they much resemble both in flavour and consistence."—*Mrs. Hussey.*

Modes of Cooking Lactarius deliciosus.—"The rich gravy it produces is its chief characteristic, and hence it commends itself to make a rich gravy sauce, or as an ingredient in soups. It requires delicate cooking, for though fleshy it becomes tough if kept on the fire till all the juice is exuded. Baking is perhaps the best process for this agaric to pass through. It should be dressed when fresh and pulpy."—*Edwin Lees.*

Stewed Deliciosus.—"The *tourtière* (or pie-dish) method of cooking suits *Lactarius deliciosus* best, as it is firm and crisp in substance. Be careful to use only sound specimens. Reduce them by cutting across to one uniform bulk. Place the pieces in a pie-dish, with a little pepper and salt, and a small piece of butter on each side of every slice. Tie a paper over the dish, and bake gently for three-quarters of an hour. Serve them up in the same hot dish."—*Mrs. Hussey.*

Deliciosus Pie.—Pepper and salt slices of the agaric, and place them in layers with thin slices of fresh bacon, until a small pie-dish is full; cover with a crust of pastry or mashed potatoes, and bake gently for three-quarters of an hour. If with potato crust, brown nicely before a quick fire.

Deliciosus Pudding.—Cut the agaric into small pieces; add similar pieces of bacon, pepper, and salt, and a little garlic or spice; surround with crust, and boil three-quarters of an hour.

Fried Deliciosus.—Fry in slices, properly seasoned with butter, or bacon and gravy; and serve up hot with sippets of toast. A steak in addition is a great improvement.

Morchella esculenta (the Morel)

Every one knows the Morel—that expensive luxury which the rich are content to procure at great cost from our Italian warehouses, and the poor are fain to do without. It is less generally known that this fungus, though by no means so common with us as some others (a circumstance partly attributable to the prevailing ignorance as to when and where to look for it, or even of its being indigenous to England), occurs not unfrequently in our orchards and woods, towards the beginning of summer. Roques reports favourably of some specimens sent to him by the Duke of Athol; and others, from different parts of the country, occasionally find their way into Covent Garden Market. The genus *Morchella* comprises very few species, and they are all good to eat. Persoon remarks, that though the Morel rarely appears in a sandy soil, preferring a calcareous or argillaceous ground, it frequently springs up on sites where charcoal has been burnt, or where cinders have been thrown.

Figure 37 *Morchella esculenta* (the Morel). Woods, & c., in the spring; colour pale buff; height, 3 to 5 inches.

Pileus very various in shape and hue, the surface broken-up into very little cells, made by folds or plaits of the hymenium, which are more or less salient, and constitute the so-called ribs. These *ribs* are very irregular, and anastomose with each other throughout; the pileus hollow, opening into the irregular stem. *Spores* pale yellow. Neither of these funguses should be gathered after rain, as they are then insipid and soon spoil.

"M. Roques says the Morel may be dressed in a variety of ways, both fresh and dry, with butter or in oil, *au gras* or *à la crême*. The following receipts for cooking them are from Persoon. 1st. Having washed and cleansed them from the earth which is apt to collect between the plaits, dry thoroughly in a napkin, and put them into a saucepan with pepper, salt, and parsley, adding or not a piece of ham; stew for an hour, pouring in occasionally a little broth to prevent burning; when sufficiently done, bind with the yolk of two or three eggs, and serve on buttered toast. 2nd. *Morelles à l'Italienne.*—Having washed and dried, divide them across, put them on the fire with some parsley, scallion, chervil, burnet, tarragon, chives, a little salt, and two spoonfuls of fine oil. Stew till the juice runs out, then thicken with a little flour; serve with bread-crumbs and a squeeze of lemon. 3rd. *Stuffed Morels.*—Choose the freshest and whitest morels, open the stalk at the bottom, wash and wipe them well, fill with veal stuffing, anchovy, or any rich *farce* you please, securing the ends, and dressing between thin slices of bacon; serve with a sauce like the last."—*Badham.*

Hygrophorus pratensis

"*Pileus* convexo-plane, then turbinate, smooth, moist; disc compact, gibbous; margin thin; *stem* stuffed, even, attenuated downwards; *gills* deeply decurrent, arcuate, thick, distant."—*Grev. t. 91; Huss. II. t. 40.*

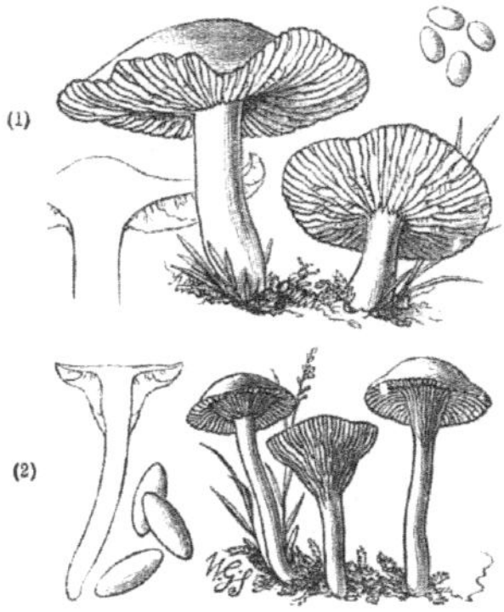

Figure 38 (1). *Hygrophorus pratensis*. Pastures, in autumn; colour, full buff; diameter, 2 to 3 inches.

Figure 38 (2). *Hygrophorus virgineus* (Viscid White Mushroom). Pastures, in autumn; snow-white; diameter, ½ inch to 1½ inches.

"On downs and short pastures. Very common. *Pileus* tawny or deep buff, sometimes nearly white, as in the next. Probably esculent."—*Berkeley.*

Hygrophorus virgineus (Viscid White Mushroom)

"*Pileus* fleshy, convexo-plane, obtuse, moist, at length areolato-rimose; *stem* stuffed, firm, short, attenuated at the base; *gills* decurrent, distant, rather thick."—*Grev. t. 166.* "On downs and short pastures. Extremely common. Mostly pure ivory-white."—*Berkeley.*

This species, exquisite in form and flavour, is one of the prettiest ornaments of our lawns, downs, and short pastures at the fall of the year. In these situations it may be found in every part of the kingdom. It is essentially *waxy*, and feels and looks precisely as if made of the purest virgin wax. The stem is firm, stuffed, and attenuated, and the gills singularly distant from each other; it changes colour a little when getting old, at which time it is unfit for culinary purposes.

A batch of fresh specimens, broiled or stewed with taste and care, will prove agreeable, succulent, and flavorous eating, and may sometimes be obtained when other species are not forthcoming.

"Several allied species enjoy the reputation of being esculent, notably *H. niveus*; and my friend Mr. F. C. Penrose has eaten, and speaks favourably of *H. psittacinus*—a highly ornamental yellow species, with a green stem, sometimes common enough in rich pastures (and *said* to be very suspicious)."—*W. G. Smith.*

Cantharellus cibarius (Chantarelle)

Figure 39 *Cantharellus cibarius* (Chantarelle). Woods, autumn; rich golden yellow; diameter, 2 to 4 inches.

When young its *stalk* is tough, white, and solid; but as it grows this becomes hollow and presently changes to yellow; tapering below, it is effused into the substance of the *pileus*, which is of the same colour with it. The *pileus* is lobed, and irregular in shape; its margin at first deeply involute, afterwards when expanded, wavy. The *veins* or plaits are thick, subdistant, much sinuated, running some way down the stalk. The *flesh* is white, fibrous, dense, "having the odour of apricots" (*Purton*) or of "plums" (*Vitt.*). "The *colour* yellow, like that of the yolk of eggs, is deeper on the under surface; when raw it has the pungent taste of pepper: the *spores*, which are elliptic, are of a pallid ochre colour." (*Vitt.*) The Chantarelle grows sometimes sporadically, sometimes in circles or segments of a circle, and may be found from June to October. At first it assumes the shape of a minute cone: next, in consequence of the rolling in of the margin, the pileus is almost spherical, but as this unfolds it becomes hemispherical, then flat, at length irregular and depressed.

"This fungus," observes Vittadini, "being rather dry and tough by nature, requires a considerable quantity of fluid sauce to cook it properly." "The common people in Italy dry or pickle, or keep it in oil for winter use. Perhaps the best ways of dressing the Chantarelle are to stew or mince it by itself, or to combine it with meat or with other funguses. It requires to be gently stewed, and a long time to make it tender; but by soaking it in milk the night before, less cooking will be requisite."—*Badham.*

Hydnum repandum (Hedgehog, or Spine-bearing Mushroom)

Figure 40 *Hydnum repandum* (Spine-bearing Mushroom). Woods, autumn; colour, pale buff; diameter, 2 to 5 inches.

Pileus smooth, irregular in shape, depressed in the centre, more or less lobed, and generally placed irregularly on the stem (eccentric); of a pale buff or cinnamon colour; from two to five inches in diameter. Flesh firm and white; when bruised it turns slightly brown. *Spines* crowded, awl-shaped, slanting, soft and brittle, varying in size and length, and of a faint cinnamon tint. *Stem* white, short, solid, crooked, and often lateral.

There is no possibility of mistaking the hedgehog mushroom: when once seen it is always to be remembered. Its awl-shaped spines are crowded beneath the pileus; its size and colour are most marked; it resembles closely, as has been said, a lightly-baked cracknel biscuit in colour.

"This fungus occurs principally in woods, and especially in those of pine and oak; sometimes solitary, but more frequently in company and in rings."—*Badham.*

Opinions on the Merits of Hydnum repandum as an Edible Fungus.—"The general use of this fungus throughout France, Italy, and Germany, leaves no room for doubt as to its good qualities."—*Roques.*

"When well stewed it is an excellent dish, with a slight flavour of oysters. It makes also a very good *purée.—Dr. Badham.*

"A most excellent fungus, but it requires a little caution in preparation for the table. It should be previously steeped in hot water and well drained in a cloth; in which case there is certainly not a more excellent fungus."—*Berkeley.*

"A wholesome fungus and not to be despised; but not in the first class as to flavour, requiring the help of condiments. It has the advantage, however, of growing later than most funguses, and may be found up to the middle of November."—*Edwin Lees.*

"One of the most excellent fungi that grows; its flavour very strongly resembles oysters."—*The Rev. W. Houghton.*

Modes of Cooking Hydnum repandum.—The hedgehog mushroom is dense in structure, and in whatever way it may be cooked, all authorities agree that it must be done slowly at a low temperature until it is tender, and with plenty of stock or white sauce to supply its deficiency in moisture.

Stewed Hydnum.—"Cut the mushrooms in pieces and steep for twenty minutes in warm water; then place in a pan with butter, pepper, salt, and parsley; add beef or other gravy, and simmer for an hour."—*Trans. from M. Roques.*

"Stew in a brown or white sauce."—*Mrs. Hussey.*

"Cut up in bits about the size of a bean, and stew in white sauce, when it will almost pass off as oyster sauce."—*The Rev. W. Houghton, F.L.S.*

Agaricus orcella (Orgelle or Vegetable Sweetbread)

Pileus thin, irregular, depressed in the centre, lobed, with undulated borders, from two to three inches across. In colour clear white, sometimes tinted with pale brown on its prominences, and occasionally with a grey centre or even lightly zoned with grey. Its surface is soft and smooth to the touch, except in wet weather, when it becomes soft and sticky. The flesh is soft, colourless, and unchangeable. *Gills* crowded, decurrent, at first nearly white, then pinkish grey, taking at length a light brown tint. Spores pale brown. *Stem* smooth, solid, short, decreasing in size; central when young, but becoming eccentric from the pileus growing irregularly. *Odour* pleasant, usually compared to that of fresh meal, but Dr. Badham and others think it resembles more closely the smell of cucumber or syringa leaf.

Figure 41 (1) *Agaricus orcella* and (2) *Agaricus prunulus* (Plum Mushroom). Woody places, in autumn; colour, snow-white, with pale rose gills; diameter, 2 to 4 inches.

Agaricus prunulus (Plum Mushroom)

Pileus fleshy, compact, at first convex, then expanded, becoming depressed in the centre, irregularly waved, and slightly pruinose; from two to five inches broad; surface dry, soft, white, or sometimes grey. The flesh thick, white, and unchangeable. *Gills* crowded, deeply decurrent, at first white, then a pale dull flesh-colour, or yellowish brown. Spores pale brown. *Stem* white, solid, firm, slightly ventricose, an inch or more long, and half an inch thick; naked, often striate, and villose at the base; often eccentric. *Odour* like that of new meal, but usually too strong to be agreeable.

There has been considerable confusion, writes Dr. Bull, between the two Agarics *orcella* and *prunulus*; some thinking that we have only *orcella* in England (*Dr. Badham*); and others only *prunulus* (the *Rev. M. J. Berkeley*), and others again that they are both the same fungus, differing only in size. Dr. Badham and some others again confuse *prunulus* with *gambosus*, the fungus of early spring, and this has arisen from the French term *mousseron* being often applied to both these funguses; but they are so essentially different as not to be liable in any way to be mistaken for each other. *Agaricus orcella* and *A. prunulus* are both placed on the same page in the illustration, so that their close alliance may be seen at a glance. Fries treats them as separate funguses, "in deference to ancient authority, since their differences are chiefly in degree." These differences are, nevertheless, so well marked, that they are kept separate here. *Orcella* is a smaller and more delicate fungus than*prunulus*. It is thinner and less fleshy, more undulated in its borders, and has a lighter and more agreeable odour. *Orcella* grows in more open glades than *prunulus*; it is usually much whiter in colour, sometimes in high situations white and glazed as an egg-shell, or even pottery. *Orcella* grows more solitary than *prunulus*, in light, scattered groups, showing an inclination for the neighbourhood of oak-trees, and where it does grow it may be found year after year in the same place, but seldom more than two or three in a spot. Last year, 1869, when *orcella* was pretty plentiful, *prunulus* was not to be found in the situations where it grows usually most abundantly. *Prunulus* is the reverse of all this. It prefers more shaded places, is larger, more fleshy, and with a strong odour rather heavy and overpowering. It grows in greater quantities together, and not unfrequently in crowded rings from four to six feet in diameter.

As edible funguses they should certainly be kept distinct. *Orcella* is light and pleasant in odour, and excellent in flavour: it is so tender and delicate as to be termed, not inaptly, "vegetable sweet-bread." *Prunulus*, on the other hand, though always good, is to many people too strong in odour, and more coarse in taste.

Opinions on the Merits of Agaricus orcella and A. prunulus.—"A very delicate mushroom."—*Dr. Badham.* "The flavour of *orcella* is very delicate, and equal to anything amongst fungi, or rather superior to the majority. The same remarks apply to *prunulus*, which I think is the same thing. It belongs to the first rank of edible fungi."—*Edwin Lees.*

Modes of Cooking Agaricus orcella and Agaricus prunulus.—*Orcella* being usually found in small quantities, is best, perhaps, when broiled and served on hot toast. *Prunulus* will yield an abundance for broiling or stewing, or both. "*Orcella*should be eaten the day it is gathered, either stewed, broiled, or fried with egg and bread-crumbs like cutlets."—*Dr. Badham.*"However prepared, it is most excellent; the flesh is firm and juicy, and full of flavour, and whether broiled or stewed, it is a most delicious morsel."—*Worthington G. Smith.* "*Orcella* will dry, and may be preserved in this way. It loses much of its volume, but it acquires *un aroma suavissimo.*"—*Vittadini. From the Transactions of the Woolhope Naturalists' Field Club.*

Edible Fungi in America.—To give an idea of the rich stores of fungi that spring up in some distant parts of the earth, and in climes so different to ours that one would at first sight suppose such fragile and fugacious bodies as fungi would not abound in them, the following interesting communication from Dr. Curtis, of South Carolina, to the Rev. W. Berkeley is here given. It will prove well worthy the attention of American readers:—

"You have asked me to give you my 'experience with the eatable mushrooms of America.' This will be most satisfactorily done, I presume, in pretty much the same style in which I would narrate it to you at your own fireside. My experience runs back only about twelve or fifteen years. You may remember that previous to this period I expressed a fear of these edibles, as I had grown up with the common prejudices against them entertained by most people in this country. Having occasionally read of fearful accidents from their use, and there being abundance of other and wholesome food obtainable, I felt no inclination to run any risks in needlessly enlarging my bill of fare. Thus I had passed middle life without having once even tasted a mushroom.

"But as under your guidance and assistance my knowledge of fungi increased, a confidence in my ability to discriminate species grew up with it, and a curiosity to test the qualities of these much-lauded articles got the better of timidity; and now, I suppose, I can safely say, that I have eaten a greater variety of mushrooms than anyone on the American continent. I have even introduced several species before untried and unknown. From the beginning of my experiments, however, I have exercised great caution, even with species long recognised as safe and wholesome. In every case I began with only a single mouthful. No ill effect following, I made a second essay upon two or three mouthfuls, and so on gradually until I made a full meal of them. Fortunately, I have never blundered upon any kind that was mischievous, although I have eaten freely of forty species. This is due, perhaps, to my general acquaintance with species that have been long used in Europe, and hence I have made no experiments upon new species which had not some affinity or analogy with them.

"For instance, *A. campestris* and *A. arvensis* being wholesome, I did not doubt but that *A. amygdalinus* (a new species closely allied to *A. arvensis*) might be safely attempted, and it has proved equally safe and palatable. Indeed, this may be regarded as the safest of all species for gathering, as it can be discriminated from all others even by a child or a blind person. Its taste and odour are so very like those of peach kernels or bitter almonds, that almost invariably the resemblance is immediately mentioned by those who taste it crude for the first time. This flavour is lost by cooking, unless the mushroom be underdone. When thoroughly cooked I cannot myself distinguish it from *A. campestris*. One or two persons have expressed the opinion that they can distinguish it, and that it is not quite so good. Others, again, are equally positive that it is better. In the crude state I deem it the most palatable of all mushrooms, as it leaves a very agreeable aftertaste upon the palate, fully equal to that of almonds. This is the thing I sent you some years since for cultivation, but which failed to grow. I very much wish it might be propagated in England, so that we might ascertain whether it would undergo any change of qualities in a different soil and climate. I have for some time been entertaining the suspicion that such is the case with many of our species. Thus, in European books the Morel is described as possessing a peculiar flavour, that has given its name to the Morello cherry. I can detect nothing of the sort in our morel. You speak of *A. Cæsareus* (in *Introd. Crypt. Bot.*) as being 'perhaps the most delicious of all fungi.' This grows in great quantities in our oak-forests, and may be obtained by the cartload in its season; but to my taste, and that of all my family, it is the most unpalatable of all our fungi, nor can I find many of our most passionate mycophagists who will avow that they like it. I have tried it in almost every mode of cookery, but without success. There is a disagreeable saline flavour that we cannot remove nor overlay.

"In the *Tricholoma* section, in which are several species long known as edible, I did not hesitate to experiment upon any that had the odour and taste of fresh flour. I began with *A. frumentaceus*, not learning from books whether it had been eaten in Europe. To this I subsequently added three new American species belonging to the same group. All are excellent when stewed, and are especially valuable for their appearance in late autumn, even during hard frosts, when other agarics are mostly out of season.

"Again, there seemed such a similarity of texture and habit between *A. cæspitosus* (*Lentinus*, Berk.) and *A. melleus*, although the former belongs to *Clitocybe*, that the temptation to a trial of it was irresistible. As it is found here in enormous quantities, and a single cluster will often contain fifty to a hundred stems, it might well be deemed a valuable species in a time of scarcity. It would not be highly esteemed where other and better sorts can be had; but it is generally preferred to *A. melleus*. I have found this species very suitable for drying for winter use.

"Among the *Boleti* I ventured, in ignorance if it had ever been eaten, to try *B. collinitus*, on account of its close relationship with *B. flavidus*. I am not particularly fond of *Boleti*, but this species has been pronounced delicious by some to whom I have sent it.

"So among the *Polypores*, I had no fear of harm from the use of a new American species (*P. poripes*, Fr.), on account of its relation to *P. ovinus*, in its texture and its flavour. The taste of the crude specimen is like that of the best chestnuts or filberts. It has been compared even with the cocoa-nut, and is certainly of very agreeable flavour. It does not, however, make a superior dish for the table, being rather too dry, but it is innocent and probably nutritious.

"Of the '*Merisma*' group of *Polypores*, having already tried *P. frondosus*, *P. confluens*, and *P. sulfureus*, I ventured, after some hesitation, and with more than usual caution, to test the virtues of a new American species (*P. Berkelei*, Fr.), notwithstanding the intense pungency of the raw material, which bites as fiercely as *Lactarius piperatus*. When young, and before the pores are visible, the substance is quite crisp and brittle, and in this state I have eaten it with impunity and with satisfaction, its pungency being all dissipated by stewing. I do not, however, deem it comparable with *P. confluens*, which is rather a favourite with me, as it is with some others to whom I have introduced it. *P. sulfureus* is just tolerable; safe, but not to be coveted when one can get better. When I say safe, I mean not poisonous. I cannot recommend it as a diet for weak stomachs, which should be said of some other fungi of similar texture. I am here reminded of an experience I had three or four years ago with this species, which would have greatly alarmed me had it happened at an earlier date in my experiments, and which would probably have deterred anyone unused to this kind of diet from ever indulging in it again. I had a sumptuous dish of it on my supper-table, of which most of my family, as well as a guest staying with us, partook very freely. During the night I became exceedingly sick, and was not relieved until relieved of my supper. My first thought on the accession of my illness was of *Polyporus sulfureus*; but as I remembered that inflammation was one of the symptoms of fungus-poisoning, and I could detect no indications of this in my case, I soon dismissed the rising fear, did not send for the doctor, nor take any remedy. Others, who had partaken of the fungus more freely than myself were not at all affected; and I presume my sickness was no more induced by the *Polyporus* than by the bread and butter I had eaten. And yet, had I alone partaken of the dish, or had one or two others been affected in like manner, doubtless the night attack would have been very confidently attributed by some to the mushroom; or had this been my first trial of that article, possibly I might ever after have regarded it with suspicion. I learned a few days afterwards from one of our physicians, that this kind of sickness was then somewhat prevalent in the community, and could be attributed to no known cause. For the credit of this species, therefore, we were fortunately able to distinguish the *post hoc* from the *propter hoc.*

"There are families in America that for generations have freely and annually eaten mushrooms, preserving a habit brought from Europe by their ancestors. In no case have I heard of an accident among them. I have known no instance of mushroom-poisoning in this country, except where the victims rashly ventured upon the experiment without knowing one species from another. Among the families above mentioned, I have not met with any whose knowledge of mushrooms extended beyond the common species (*A. campestris*), called pink gill in this country. Several such families live near me, but not one of them was aware, until I informed them, that there are other edible kinds. Everything but the pink gill, which had the form of a mushroom, was to them a toadstool, and poisonous. When I first sent my son with a fine basket of Imperials (*A. Cæsareus*), to an intelligent physician, who was extravagantly fond of the common mushroom, the lad was greeted with the indignant exclamation, 'Boy, I wouldn't eat one of those things to save your father's head!' When told that they were eaten at my table, he accepted them, ate them, and has eaten many a one since, with all safety and with no little relish. Since that time our mycophagists eat whatever I send them, without fear or suspicion.

"I have interested myself to extend the knowledge of these things among the lovers of mushrooms, and also their use among those who have not before tried them. In the latter work I am not always successful, on account of a strong prejudice against vegetables with such contemptible names, and an unconquerable fear of accidents. Yet, as in my own case, curiosity often conquers these errors. When away from home I have frequently obtained permission from a kind hostess to have cooked a dish of mushrooms that I have found on her premises. It has rarely occurred in such cases that the dish, then tasted for the first time, was not declared to be delicious, or the best thing ever put in the mouth. This latter phrase was once used in reference to so indifferent an article as *A. salignus*. Indeed, I have found several persons who class this amongst the most palatable species. To such persons a dish of fresh mushrooms need seldom be wanting, as this one can be had every month of the year in this latitude. I am induced to believe that the quality of this species varies with the kind of wood it grows from, and that it is better flavoured when gathered from the mulberry, and especially from the hickory, than when taken from most other trees. Its fitness for the table seems also to depend much upon the rapidity of its growth; those which grow slowly, as is the case with some of our garden vegetables, being of tougher texture and of less delicate flavour. A warm sun after heavy rains brings them out in greatest perfection.

"I have several times been asked by persons eating mushrooms for the first time, whether these things belong to the vegetable or animal kingdom. There is certainly a very noticeable resemblance in the flavour of some of them to that of flesh, fish, or mollusc, so that the question, as founded merely on taste, is not an unnatural one. But I was much struck with its propriety when reading an article in 'Fraser's Magazine,' a few years since, written by the late Mr. Broderip, who therein says that mushrooms contain osmazome. If this be so, it accounts both for their flavour and for their value as food. Of this latter quality I had become so well convinced that, during our late war, I sometimes averred, and I doubt if there was much, if any, exaggeration in the assertion, that in some parts of the country I could maintain a regiment of soldiers five months of the year upon mushrooms alone.

"This leads to a remark which should not be overlooked, upon the great abundance of eatable mushrooms in the United States. I think it is Dr. Badham who boasts of their unusual number in Great Britain, stating that there are thirty edible species in that kingdom. I cannot help thinking that this is an under-estimate. But if the Doctor is correct, there is no comparison between the number in your country and this. I have collected and eaten forty species found within two miles of my house. There are some others within this limit which I have not yet eaten. In the catalogue of the plants of North Carolina, you will notice that I have indicated one hundred and eleven species of edible fungi known to inhabit this State. I have no doubt there are forty or fifty more, as the alpine portion of the State, which is very extensive and varied, has been very little explored in search of fungi.

"In October, 1866, while on the Cumberland Mountains in Tennessee, a plateau less than 1000 feet above the valleys below, although having very little leisure for examination during the two days spent there, I counted eighteen species of edible fungi. Of the four or five species which I collected there for the table, all who partook of them, none of whom had before eaten mushrooms, most emphatically declared them delicious. On my return homeward, while stopping for a few hours at a station in Virginia, I gathered eight good species within a few hundred yards of the dépôt. And so it seems to be throughout the country. Hill and plain, mountain and valley, woods, fields, and pastures, swarm with a profusion of good, nutritious fungi, which are allowed to decay where they spring up, because people do not know how, or are afraid, to use them. By those of us who know their use their value was appreciated, as never before, during our late war, when other food, especially meat, was scarce and dear. Then such persons as I have heard express a preference for mushrooms over meat had generally no need to lack grateful food, as it was easily had for the gathering, and within easy distance of their homes if living in the country. Such was not always the case, however. I remember on one occasion during the gloomy period, when there had been a protracted drought, and fleshy fungi were to be found only in damp, shaded woods, and but few even there, I was unable to find enough of any one species for a meal; so gathering of every kind, I brought home thirteen different kinds, had them all cooked together in one grand *pot pourri*, and made an excellent supper. Among these was the Chantarelle, upon which I would say a few words in confirmation of what I have already said upon the varying qualities of mushrooms in different regions and localities. You have somewhere written of this mushroom as being so highly esteemed a delicacy, that it is much sought for when a dinner of state is given in London. Can this be because it is a rarity? (for nothing common and easily obtained is deemed a delicacy, I believe), or because you have it of finer flavour in England? Here, where it abounds, no one seems to care at all for it, and some would forego mushrooms entirely rather than eat this. It certainly varies much in quality, as I have occasionally found it quite palatable, and again, though cooked in the same mode, very indifferent. I have been unable to ascertain whether this difference is due to locality, exposure, shade, soil, moisture, or temperature. That soil has much to do with the flavour of some species of mushrooms I am well convinced. In a parcel of pink gills I have sometimes found one or two specimens, though perfectly sound, of such unpleasant odour and taste as would spoil a whole dish. So also with the snowball (*A. arvensis*), of which I annually find a few beautiful specimens growing near my residence, upon a grassy turf which covers a pile of trash made up of decomposed sticks, leaves, and scrapings from the adjoining soil. Their taste and odour are perfectly detestable. I had one specimen cooked, but no amount of seasoning could abate the offensiveness of the odious thing; yet within a hundred yards of these I gather specimens of the same identical species, which are of fine flavour, equal to that of the best mushrooms. As I have before intimated the varying flavour of mushrooms growing on different kinds of wood, so here I suppose the unpleasant qualities of some specimens of these two well-known and favourite species, may be owing to something in the soil where they grow which they cannot assimilate, and so render a palatable and wholesome species totally unfit for the table. Whether such specimens, if eaten, would be poisonous or unwholesome, I do not feel any temptation to prove. It is not probable that they will ever do any mischief, for it is incredible that any human being should so pervert his instincts as to swallow such a villanous concoction.

"Experience and observations like these would perhaps justify the inference that an innocent species may sometimes be deleterious, on account of its taking up some bad element from the soil. But as I have never known a case of poisoning in families that are well acquainted with the common mushroom or pink gill, that gather the specimens for themselves, and have used this article of food annually for many generations, I cannot agree with a suggestion somewhere made by you, that perhaps all mushrooms contain a poisonous element, but some of them in such small quantity as to have no appreciable effect. Now, had you seen the quantities of stewed mushrooms swallowed at a single meal which I have seen thus devoured, and with no more harm than from the same amount of oyster or turtle soup, I think you would be forced to the conclusion that such an amount, even of poisonous infinitesimals, must have had some very unpleasant manifestations, or else be a very innocent diet.

"It is said that the sale of the pink gill (*A. campestris*) is forbidden in the Italian markets, because that species has often proved to be poisonous. May not this have been occasioned by ignorant and careless collectors or by worthless inspectors? To us in America, who use this species so freely and fearlessly, the Italian's curse, 'May he die of a Pratiolo!' would have no more terror than 'May he die of aromatic pain.'

"Our best and standard mushrooms are the pink-gill (*A. campestris*); snowball (*A. arvensis*); peach-kernel (*A. amygdalinus*); nut (*A. procerus*); French (*A. prunulus*); morel (*M. esculenta*); coral (*Clavaria*); and omelette (*Lycoperdon giganteum*). These are almost universally in high esteem. Yet tastes differ on these things as on fruits and vegetables; some putting one, some another, at the head of the list, though fond of all and ever ready to use any of them—as one who prefers a peach may yet relish an apple. There are some among us who regard *A. procerus* as fully equal to *A. campestris*, and I am almost of the same opinion. When broiled or fried it truly makes a luscious morsel. I mention in this connexion, that this species here bears the name of nut mushroom, from a quality that I do not find mentioned in the books which describe it. The stem when fresh and young has a sweet nutty flavour, very similar to that of the hazel nut. Is this the case with you? Its flavour is so agreeable that I am fond of chewing the fresh stems. From this peculiarity in connexion with its movable ring, its form and colours, I deem it a perfectly safe species to recommend for collecting. We have no species likely to be mistaken for it, except *A. rachodes*, and I fully tested the innocence of this before commending the first to others. This has been suspected by some, but I have found it harmless. Though pretty well flavoured, it is not comparable with *A. procerus*, and the flesh is so thin and spongy that no one would choose it when those of more compact texture are to be had. *A. excoriatus*, of the same group, is a much preferable species.

"The Morel is one of my greatest favourites, but this is not found in quantity except in calcareous districts. A few days since (April 21) I had a dozen for supper, the largest number I ever had at one time.

"The *Lycoperdon giganteum* is also a great favourite with me, as it is, indeed, with all my acquaintances who have tried it. It has not the high aroma of some others, but it has a delicacy of flavour that makes it superior to any omelette I have ever eaten. It seems, furthermore, to be so digestible as to adapt it to the most delicate stomachs. This is the South Down of mushrooms.

"In this latitude (about 36 degrees) we can find good mushrooms for the table during nine or ten months of the year. Including *A. salignus*, which some are quite fond of, we can have them in every month, as this species comes out during any warm spell in winter. *A. campestris* makes its appearance here as early as March, but is not in full crop until September. Several excellent species of the *Tricholoma* group do not spring up until after frost sets in, and continue into December. Such is the case too with *Boletus collinitus*, which sometimes emerges from the earth frozen solid.

"These observations and experiences are confined chiefly to the Carolinas; though I presume, from casual observations elsewhere, and from information derived from correspondents in other States, that, making some allowance for difference of climate and length of seasons, what I have said is generally applicable to the whole country."

Why we should not eat Funguses

The following interesting paper from the Rev. J. D. La Touche was read at a meeting of the Woolhope Naturalists' Field Club:—

"It is said that at Rome, when a mortal is about to be raised to the dignity of sainthood, the precaution is taken of providing a 'devil's advocate,' who, by pointing out as strongly as he can all the faults of the candidate, secures the fair discussion of both sides of the question, and is a guarantee, moreover, that no unworthy aspirant to such exalted honours should be rashly admitted to them.

"On the present occasion I make bold to present myself in this unamiable capacity. No member, indeed, of this respected Club is seeking canonization, yet, a step not less important is contemplated in the enrolment of a hitherto despised and even abhorred member of the vegetable kingdom among the list of its edible products; indeed, some may consider such a step as of more importance to our race than the apotheosis of a peccant mortal; and therefore it would appear that, if in the one instance it is desirable that all the peccadilloes of the candidate should be exposed, *a fortiori*, it must be so in the other.

"Let me, then, first observe that these gentlemen at the bar have actually a very bad character, and that it is not likely that this would be the case unless they were really great sinners.

"Here, some will exclaim, no doubt, 'Prejudice, my dear sir! vulgar prejudice is capable of the grossest injustice—ignorant prejudice has driven from our tables a delicious article of food, and deprived the poor of a wholesome diet.' It is often said that he was a brave man who first ate an oyster, and truly a more uninviting mouthful than it was could scarcely be imagined; and yet the fact that it *is* good and wholesome soon disposed of any prejudice against it. And is it not likely that such would be the case, were the fungus tribe fit for human food? Can we suppose any prejudice arising from their leathery looks would not evaporate like mists before the morning sun, were they really the nutritious and delicious dainties they are described to be by their enthusiastic advocates?

"I think it may be observed that the general character which a man bears is, on the whole, a true one. That big school, the world in which we live, contrives, in some way or other, to hit off pretty accurately our average merit and take our measure, and though it may make a mistake now and then in some particular instance, its general estimate is a fair one; and so with funguses. There may be a too-sweeping condemnation of all kinds of them: nay, it may be even probable that *Agaricus campestris* is not the best that grows, and yet, after all, the prevalent distrust of the tribe is well founded.

"When, *e.g.*, some family in a parish is known to have been poisoned by eating a wrong sort, it is not surprising, nor can it be called stupid prejudice, if their neighbours are ever after rather shy of the article of food which produced that result. But it will be said that the mischief arose from ignorance—had that family known the marks that distinguish between the wholesome and the poisonous kinds this would never have taken place. If ever there was a case in which ignorance was bliss, surely this is it. A short time ago, I accompanied a scientific friend in a foray among the funguses, which we made with a special view to the improvement of our intended repast, and was on that occasion struck with the elaborate precautions which seemed to be necessary to observe in discriminating the good from the bad. It would almost seem that Nature had purposely contrived a labyrinth of ingenious stumbling-blocks to guard this mysterious product from the insatiable appetites of mankind; and so it came to pass after all, my good friend—who really seemed well up in the subject, and who found at every turn some well-known test of wholesomeness or otherwise to guide him in the specimens we collected—wound up the day by nearly poisoning a member of my family: for he had, it appears, mistaken *Boletus flavus*, a violent poison, for the very similar but wholesome and excellent *Boletus luteus*—the only difference being that the pores of the one are somewhat smaller and less angular than those of the other. Surely, in this instance, knowledge (and it was not in his case a little knowledge either) was a dangerous thing.

"But still it may be said that there are species the characters of which are sufficiently well-defined, and that from these, at least, the stigma ought to be removed. But even so, I would submit one or two questions to those who may be inclined to admit this. 1st. Is it so clear that a fungus which agrees with one person may not be very injurious to another? One man has, to use a vulgar expression, the stomach of a horse. Can I, an average mortal, calculate on possessing such a treasure? I saw with my own eyes my scientific friend eat and swallow an entire *Boletus flavus*, raw, without any apparent bad effects either that evening or the following day, whereas a small portion of the same kind, cooked too (I cannot, however, say *secundum artem*), produced violent sickness on another individual, who, moreover, had never before experienced sickness; indeed, this fact would seem to suggest that the stomach may be 'educated' by long habit to bear this noxious food, and, therefore, that its evil effects (harmless upon organs well trained) happen when the *experimentum in corpore vili* is tried. My friend assures me that he has eaten the highly poisonous *Boletus satanas* with no worse effect than a little indigestion the next morning. Can, I would ask, the experience of such a seasoned digestive apparatus as his be any guide to those who have not gone through the course of training which he has?

"Again, may it not be possible that the same kind of fungus which in some instances is wholesome, may, if grown under different circumstances, and supplied with different nutriment, assume very different properties? And again, are we competent to judge of the wholesomeness of a particular article of food unless it is tried by a very large number of persons—unless it be 'exhibited,' to use a medical term, on a great variety of constitutions? Indeed, is there not some ground for thinking that such an exhibition would be in many instances far from satisfactory?

"On the whole, it would appear that the advice of an eminent physician, an ardent admirer of the fungus, was good and sound. When he heard of the escape my family had on this occasion, he said that this article of diet should be partaken of with 'great caution.' And by the way, is not this itself a very suspicious expression? 'Great caution!' If I am introduced to a gentleman, and told at the same time that I must conduct myself towards him with 'great caution' or he will probably do me some deadly mischief, it would hardly be thought a very hearty and promising introduction; yet here we are told that this excellent family to which we are so warmly introduced has some members belonging to it so villanously disposed, that possibly we may pay for our acquaintance with our lives. This is not very encouraging, and so the course adopted by a young lady who indulges in these experiments, to whom I was speaking the other day, would seem to be a very prudent one. She says she never partakes of these dainties till she has seen the effect they have had upon somebody else! But even so, only picture the ghastly scene which a banquet of this kind would present; each guest looking anxiously into his neighbour's face, awaiting in terror the contortions which are to show that he has partaken of the fatal dish."

While Mr. La Touche's paper should not deter us from using and showing others the value of the quantities of *edible*fungi now generally allowed to rot in our fields and woods, and nowhere perhaps so abundant as in the pleasure grounds and woods round country seats, yet, as impressing the necessity of using due discrimination in gathering, it may be read with advantage by all.

Made in the USA
Middletown, DE
05 November 2026

36544996R10081